ANIMAL ALGORITHMS

ANIMAL ALGORITHMS

EVOLUTION AND THE MYSTERIOUS ORIGIN OF INGENIOUS INSTINCTS

ERIC CASSELL

SEATTLE   DISCOVERY INSTITUTE PRESS   2021

Description

How do some birds, turtles, and insects possess navigational abilities that rival the best manmade navigational technologies? Who or what taught the honey bee its dance, or its hive mates how to read the complex message of the dance? How do blind mound-building termites master passive heating and cooling strategies that dazzle skilled human architects? In *The Origin of Species* Charles Darwin conceded that such instincts are "so wonderful" that the mystery of their origin would strike many "as a difficulty sufficient to overthrow my whole theory." In *Animal Algorithms*, Eric Cassell surveys recent evidence and concludes that the difficulty remains, and indeed, is a far more potent challenge to evolutionary theory than Darwin imagined.

Library Cataloging Data

Animal Algorithms: Evolution and the Mysterious Origin of Ingenious Instincts by Eric Cassell

Library of Congress Control Number: 2021947213

ISBN: 978-1-63712-006-4 (paperback), 978-1-63712-007-1 (Kindle), 978-1-63712-008-8 (EPUB)

BISAC: SCI070060 SCIENCE/Life Sciences/Zoology/Ethology (Animal Behavior)

BISAC: SCI027000 SCIENCE/Life Sciences/Evolution

BISAC: SCI075000 SCIENCE/Philosophy & Social Aspects

Publisher Information

Discovery Institute Press, 208 Columbia Street, Seattle, WA 98104

Internet: http://www.discoveryinstitutepress.com/

Published in the United States of America on acid-free paper.

First Edition, First Printing, November 2021.

ADVANCE PRAISE

This new book fills an important gap in the literature about problems for neo-Darwinism and empirical evidence for intelligent design theory. While there are many works on highly complex systems in genetics, molecular machines, and anatomy, this work focuses on the utterly mysterious origin of complex programmed animal behavior and instincts. From the navigation of migrating birds and butterflies to the dance of honey bees and the miraculous abilities of other insect societies there are abundant phenomena for which Darwinism has failed to provide any plausible and adequate explanation, while the obvious design explanation has been excluded a priori by mainstream academia. This book is another welcome and highly recommended step towards an overdue paradigm change in modern biology.

—**Günter Bechly**, PhD, paleoentomologist, senior fellow with the Center for Science and Culture, former curator in the department of paleontology for the State Museum of Natural History in Stuttgart, Germany

All computer programs are algorithms. Eric Cassell wonderfully describes the clever algorithms in many animals. Where did these embedded computer programs come from? Specified complexity, irreducible complexity, and the Cambrian explosion are inexplicable from a Darwinian viewpoint. In this book, Cassell masterfully adds animal algorithms to the list.

—**Robert J. Marks II**, PhD, distinguished professor of electrical & computer engineering, Baylor University; FIEEE and FOSA; director of the Walter Bradley Center for Natural & Artificial Intelligence; editor-in-chief, *BIO-Complexity*

Eric Cassell has given us an extremely well-researched book that is enjoyable to read and addresses what is surely the most fascinating aspect of animals: their behavior. Examples of some intriguingly complex behaviors are clearly explained, along with the ways that the traditional evolutionary paradigm has struggled to explain their origin. Cassell shows that some complex animal behaviors appear to be based on programs. But if so, then who or what programmed them? *Animal Algorithms* reveals how this is a particularly difficult question for Darwinian evolution to answer. In contrast, intelligent agents are the only sort of cause known to be able to generate computer algorithms. Perhaps, then, intelligent design played a role in the origin of these complex programmed behaviors we find in these animals. Hopefully, this book will awaken more minds among the general public, and the scientific community, to possibilities beyond the rapidly aging creed of Darwinism.

—**Malcolm Chisholm**, PhD, entomology, Bristol University; MA, zoology, Oxford University; president of Data Millennium LLC; recipient, Data Management Association International Lifetime Achievement Award

As a research ecologist, I appreciate Eric Cassell's thorough and readable exploration of complex programmed behavior in animals from an engineer's perspective. Well researched with concise explanations and examples, *Animal Algorithms* employs a systems biology approach to ably expose neo-Darwinian mechanisms as deficient and show that design inferences offer tenable explanations that point to multiple vistas of further scientific investigation. A stellar contribution that extends the breadth of ID research.

—**George A. Damoff**, PhD, adjunct research faculty, Division of Environmental Science, Arthur Temple College of Forestry and Agriculture, Stephen F. Austin State University

Animal Algorithms offers a fascinating exploration of some of the astonishingly complex programmed behaviors exhibited in the animal kingdom. In wonderfully concise and accessible language, Cassell explains the intricacies of these behaviors and builds a compelling argument for their intelligent design. Readers will come away with a clear understanding of why the algorithmic dances of organisms such as bees, ants, and butterflies pose an enormous challenge to the materialist evolutionary paradigm. This book is a true contribution to the ongoing conversation.

—**Melissa Cain Travis**, PhD, author of *Science and the Mind of the Maker*; affiliate faculty, Colorado Christian University

Eric Cassell has asked some very important questions about the origins of complex behavioral patterns in animals. The coordination of anatomy, physiology, communication, etc. to produce complex behavior is at least partially understood in many animals, but the origin of these behaviors is a huge mystery. The remarkable migration of monarch butterflies or the eusocial behaviors of ant colonies have long fascinated humanity with little explanation of how they came about. Are genes sufficient to account for new behaviors? Cassell suggests that intelligent design provides a far better framework for understanding the origin of these and other astounding behaviors than methodological naturalism. *Animal Algorithms* breaks new ground here in a thought-provoking discussion. Well worth the read!

—**Bruce Evans**, PhD neurobiology, Emory University; professor of biology, department chair, Huntington University

DEDICATION

In memory of my mother and father for
their continuous love and support.

ACKNOWLEDGMENTS

I want to thank the Discovery Institute for their willingness to take on the project to publish this work. In particular, I appreciate the support and encouragement provided by John West. Thanks also to Jonathan Witt for his outstanding guidance in editing the manuscript, and to Casey Luskin for a number of helpful suggestions.

CONTENTS

1. Genius in Lilliput

Zoologists have engaged in such extreme denial of motivation and goal-directed behavior, not to mention animal consciousness and complex intellectual abilities, that until very recently mechanisms for them are not widely sought or even hypothesized. At present, this is perhaps the greatest conceptual void in evolutionary ethology.[1]

— Mary Jane West-Eberhard

THERE IS GENIUS IN LILLIPUT. I DON'T MEAN JONATHAN SWIFT'S Lilliput, a fictional island peopled with petty humans six inches tall. I mean the Lilliputian world of birds and bees, termites, ants, and butterflies. There is genius here, and that genius poses a mystery, particularly in the case of clever insects. Their brains can be as small or smaller than a sesame seed, and yet these insects perform extraordinary mental feats.

Honey bees live in complex social communities where there is a division of labor based on a caste system. Each bee knows its assigned function and carries out its responsibilities accordingly. Honey bees are also expert navigators and communicators, helping them forage for food and locate new sites for their hives. This despite the fact that a bee brain has only one thousandth of one percent of the neurons found in the human brain.

Monarch butterflies migrate annually two thousand to three thousand miles between Canada and Mexico. It takes up to three generations of butterflies to complete the journey, suggesting the knowledge of the migration route is innate rather than learned. Each generation of monarchs has a clear goal for its segment of the annual migration. The accuracy of their navigation is such that they often spend the winter in Mexico in the same tree as their predecessors.

Figure 1.1. Monarch Butterflies

Spiderwebs are constructed of silk, which has several unique properties that human scientists struggle to replicate, including strength and elasticity. But equally remarkable is the behavior of the spiders in their spinning of the webs. The shape of the webs they engineer is elegant and exquisitely functional. And when part of a web is damaged, the spider promptly begins repairs to restore the original design. In addition to trapping prey, the webs enhance the spider's ability to locate prey once trapped. Even in the dark spiders can determine the exact location of trapped prey based on the vibrations in the web sensed through their legs.

Some species of termites construct nests that have impressed architects, engineers, and artists alike. The nests can reach more than twenty feet high and typically include a royal chamber, nurseries, gardens, waste dumps, a well, and a ventilation system that reduces heat and removes carbon dioxide.

Adult wasps feed on nectar, but they hunt for other insects to provide food for their larvae. The insects they hunt vary according to wasp species, and include honey bees, beetles, tarantulas, and cicadas. But the most amazing aspect of this is the wasp's ability to paralyze the captured prey.[2] The location of the neural ganglion that must be injected with a neurotoxic venom to paralyze the prey differs from one prey species to another.[3] For example, a wasp that specializes in honey bees "inserts her sting accurately between two distinct plates on the underside of the bee's neck, immobilizing but not killing it."[4]

Research has confirmed that the recognition of prey is innate, and that the stinging behavior, which must be done with precise accuracy to work, is controlled by a motor program—that is, a series of sub-routines ordered in a particular sequence to perform a given movement or task. And no simple one. To grasp this, imagine the software program that would be required to enable an advanced micro-drone to deliver a neurotoxin to the precise location in the honey bee to immobilize it. In assessing the complexity and evolution of this wasp behavior, Jerry Fodor and Massimo Piatelli-Palmarini conclude that "such complex, sequential, rigidly pre-programmed behaviour could have gone wrong in many ways, at any one of the steps... Such cases of elaborate innate behavioural programs cannot be accounted for by means of optimizing physio-chemical or geometric factors."[5]

The above examples of innate or programmed behaviors are just a handful of numerous such instances in the animal kingdom. Surprisingly, in many instances the behaviors of what we normally think of as primitive animals can be just as complex as those of more advanced animals, including mammals. Indeed, there is little correlation between the cognitive capacity of animals and their ability to produce sophisticated, apparently innate behaviors. The reason may be that such behaviors really are programmed and therefore innate, so the animals do not require significant cognitive capacity to perform them. What they do require is the specific neural "circuitry" that controls the behavior—circuitry that is quite sophisticated but apparently does not require large brains.

Effusive descriptions of these behaviors can be found in everything from National Geographic television programs to science books and articles. Jennifer Ackerman's *The Genius of Birds*[6] and Martin Giurfa's "The Amazing Mini-Brain: Lessons from a Honey Bee"[7] are two examples among many. The world of science is astounded by some of the complex innate behaviors found in the animal kingdom.

Many of these behaviors are routinely described as *enigmatic* or *mysterious*, because their origin is not understood. Thus do we encounter book titles such as *The Mystery of Migration*[8] and *Nature's Compass: The Mystery of Animal Navigation.*[9]

In *On the Origin of Species* the nineteenth-century naturalist Charles Darwin laid out his revolutionary case for common descent by gradual evolution. Darwin could not be faulted for timidity. He pressed his case at nearly every turn. But even he conceded at one point in the book that many instincts are "so wonderful" that their development "will probably have occurred to many readers, as a difficulty sufficient to overthrow my whole theory."[10]

Undaunted, however, he went on to insist that instincts were essential elements of his theory, and like the great variety of biological forms, they too developed through gradual evolution. "I can see no difficulty in natural selection preserving and continually accumulating variations of instinct to any extent that may be profitable," he wrote. "It is thus, as I believe, that all the most complex and wonderful instincts have originated. No complex instinct can possibly be produced through natural selection, except by the slow and gradual accumulation of numerous, slight, yet profitable, variations… The canon of 'Natura non facit saltum' applies with almost equal force to instincts as to bodily organs."[11]

Curiously, Darwin deleted the last sentence from later editions of *The Origin*, although he continued to adhere to its principle. A primary aim of the present book is to explore whether Darwin's assertion about the origin of complex instincts stands up to current evidence. Does the accumulated evidence from the intervening 160-plus years support the idea, in broad outline at least? If not, is there a better explanation—one drawn either from what is known as the "extended evolutionary synthesis" or from an explanation that reaches beyond that paradigm? This is the central question of the present book.

Complex programmed behaviors are evident throughout the animal kingdom, but in these pages the focus will primarily be on less advanced animals. The reason is that more advanced animals, such as primates, have significant cognitive ability, so they exhibit much more of a combination of programmed and learned behaviors, and in such cases the two are not always easily disentangled. It is easier to discriminate between programmed and learned behaviors in less advanced animals, such as bees and butterflies.

Explaining the origin of these programmed animal behaviors in evolutionary terms is challenging because the behaviors themselves are, in many cases, quite complex and likely undergirded by an extraordinarily sophisticated neurological substrate. Animal behaviors are also strikingly diverse, arguably just as diverse as the breathtaking diversity of physical characteristics we find in the animal kingdom. Those factors alone do not mean the explanatory task is impossible. But it does mean that something more than breezy just-so stories are required to provide a causally adequate explanation for their evolution.

Adding to the difficulty: animal behaviors leave relatively few clues in the fossil record. Whether for these or other reasons, the origin of animal behavior has not been studied or discussed in modern biology to the same degree as animal physiology and genetics. That's unfortunate since one of the most fascinating things about many animals is their behavior.

Another reason the subject merits greater attention: on some accounts, animal behavior is crucial to evolutionary theory as a whole. Ernst Mayr, a leading evolutionary theorist of the twentieth century, held that a change in behavior is the crucial factor initiating evolutionary innovation.[12] Mayr further argued that "behavioral shifts have been involved in most evolutionary innovations, hence the saying—behavior is the pacemaker of evolution."[13]

The subject, then, cries out for further investigation.

In addition to examining several different types of complex programmed behaviors in the animal kingdom, we will weigh the accumulated evidence in the light of some competing explanatory hypotheses, including neo-Darwinism, to see if any of them appear causally adequate. This involves use of a method common to the historical sciences known as "inference to the best explanation."

From Aristotle to Darwin

FORMAL STUDY of animals and animal behavior stretches back to the ancients. The fifth century BC philosopher Empedocles offered an account of the origin of animals that anticipated Darwin's idea of evolution by natural selection (though not Darwin's emphasis on gradualism). In the same century, the Greek atomists Leucippus and Democritus proposed a purely materialistic conception of life and the universe, one with an evolutionary component and little if any room for notions of purpose. But the man whose views came to dominate Western thinking for some two millennia took a very different view. Aristotle (384–322 BC), considered the father of biology, studied and documented the behavior of a variety of animals. In the *Movement of Animals* he wrote, "At the beginning of the inquiry we must postulate the principles we are accustomed constantly to use for our scientific investigation of nature, that is we must take for granted principles of this universal character which appear in all Nature's work. Of these one is that Nature creates nothing without a purpose."[14]

Thus Aristotle, unlike the atomists, viewed animals and animal behavior in teleological terms, where behaviors have specific functions and goals. That may seem commonsensical—and perhaps for that reason his view did predominate in the West for some two millennia. But as we will see, the Aristotelian position on this point fell out of favor in the modern era.

The idea that species have remained unchanged since their creation held sway in biology through the eighteenth century. Jean-Baptiste Lamarck (1744–1829) broke with that idea by proposing a theory of evolution, one that also attempted to explain the origin of animal behaviors. His *Zoological Philosophy* (originally published in 1809, the year Darwin was born) was the first significant effort at developing a comprehensive theory where all living organisms developed from primitive ancestors.

A primary element of his theory was that organisms had an inherent tendency to evolve toward increasing complexity. However, he is better known for the second element of his theory: the inheritance of acquired characteristics. The driving force behind this element is the environment. "The environment affects the shape and organisation of animals," he wrote, "that is to say that when the environment becomes very different, it produces in course of time corresponding modifications in the shape and organisation of animals."[15] He emphasized that this was an indirect effect, rather than a direct modification. This constituted his First Law. Lamarck's Second Law was that these acquired characteristics would be inherited by the animal's offspring. Lamarck believed that it was the change in habits (behavior) that induced the subsequent physical changes.[16]

Lamarck cited several examples of evolution that he claimed proved the theory of the inheritance of acquired characteristics. One example concerned the lengthening of tongues among anteaters, green woodpeckers, and hummingbirds.[17] Another was the height of giraffes, where their habit of foraging on trees is said to have lengthened their necks and legs to reach higher.[18] Lamarck's theory enjoyed much support in the 1800s. But late in the nineteenth century, August Weismann conducted an experiment wherein he cut off the tails of mice over several generations. Under Lamarckism one would expect that eventually the mice in later generations would be born with shorter tails. However, that did not occur. This and later more definitive experiments led to the eclipse of Lamarckism. Often overlooked, however, is the fact that his idea that animal behavior could drive the evolution of new physical characteristics became an important part of Darwin's own theory of evolution.

Indeed, Lamarck's influence on Darwin was broader even than this. Science historian Peter Bowler writes, "Darwin's lifelong commitment to a limited amount of Lamarckism and to what was later called blending inheritance (the mixing of parental characters) were integral parts of his worldview."[19] He applied this to both physical characteristics and behavior. Darwin, like Lamarck, believed an animal's habits influenced its physiology, which then could be inherited by its offspring. He cited several examples, including the drooping ears of some domestic animals,[20] and flightless birds on islands that lack predatory animals.[21] Darwin summarized his thinking on use and disuse thus: "On the whole I think we may conclude that habit, use, and disuse, have, in some cases, played a considerable part in the modification of the constitution, and of the structure of various organs; but that the effects of use and disuse have often been largely combined with, and sometimes overmastered by, the natural selection of innate differences."[22]

In offering a definition for the concept of instinct, Darwin wrote that an action performed by an animal, "more especially by a very young one, without any experience, and when performed by many individuals in the same way, without their knowing for what purpose it is performed, is usually said to be instinctive."[23] Darwin also viewed instincts as being analogous to habits. He applied the concept of use and disuse to behavioral instincts, writing, "I can see no difficulty, under changing conditions of life, in natural selection accumulating slight modifications of instinct to any extent, in any useful direction. In some cases habit or use and disuse have probably come into play."[24]

Turning again to domesticated animals, he cites examples of how behaviors change through humans deliberately selecting for specific behaviors.[25] Even under the assumption that habits could be inherited, Darwin did recognize the limitations in applying the principle of variation and natural selection to certain behaviors, writing, "It can be clearly shown that the most wonderful instincts with which we are acquainted, namely, those of the hive-bee and of many ants, could not possibly have been thus acquired."[26]

Darwin recognized the fundamental difference between the complex behaviors of animals and human abilities, the latter largely acquired through learning. He writes, "Man cannot, on his first trial, make a stone hatchet or a canoe... He has to learn this work by practice; a beaver, on the other hand, can make its dam or canal, and a bird its nest, as well, or nearly as well, and a spider its wonderful web, quite as well, the first time it tries as when old and experienced."[27]

Darwin knew nothing of genes and genetic mutations. But after scientists began to unravel the mystery of genetics, the insights were incorporated into Darwinism and eventually christened *the modern synthesis*, a label used by Julian Huxley, grandson of Darwin defender T. H. Huxley, in the book *Evolution: The Modern Synthesis* (1942). This updated form of Darwinism was developed based on the work of several scientists in various disciplines, including zoologist Ernst Mayr, geneticists Theodosius Dobzhansky, Ronald Fisher, Thomas Hunt Morgan, and J. B. S. Haldane, and paleontologist George Gaylord Simpson.[28]

The prevailing view established by the modern synthesis typically goes by the name neo-Darwinism, the core of which is the idea that evolution is based primarily on random genetic mutations and natural selection.[29] The synthesis is more complicated than that, however, with evolution said to occur through a number of different mechanisms. These include genetic mutation, genetic recombination, gene duplication, genetic drift, gene flow, founder effect, bottleneck effect, and others. The common thread running through all of these mechanisms is that they are centered on individual genes.

Evolutionary biologist Michael Lynch aggregates these mechanisms into four fundamental forces of evolution: natural selection, mutation, recombination, and genetic drift. "Given the century of work devoted to the study of evolution," he writes, "it is reasonable to conclude that these four broad classes encompass all of the fundamental forces of evolution."[30] He further explains that evolution cannot be understood only in terms of natural selection and adaptation. The reason is that the other forces cited are "nonadaptive in the sense that they are not a function of the fitness properties of individuals."[31]

The above represents the barest summary of modern evolutionary theory, which today includes various creative add-ons and suggested modifications. Later in the book we will consider some of the proposed ancillary mechanisms for the evolution of complex programmed behaviors. But even today many biologists regard natural selection and random genetic mutations as evolutionary theory's twin pillars, so here we will slow down to unpack each of those ideas a bit more.

Natural selection is, in very rough terms, "survival of the fittest." If, for instance, an offspring has a random genetic mutation that makes it a little faster or stronger or more clever, and if the given mutation improves its chances of surviving and reproducing, nature is thus more likely to pass that mutation down to its offspring—natural selection. On the neo-Darwinian view, a long series of such random mutations, sifted thus by natural selection, led to new forms gradually arising over the many hundreds of millions of years of evolutionary history, all beginning from one or a few original single-celled organisms. And as we saw above, Darwin made no exception for complex programmed behaviors, which he called *instinct*. He and others after him argued that these, too, could be explained by the joint mechanism of random variation, broadly understood, and natural selection.

The Zoologists vs. the Psychologists

IN THE early part of the twentieth century, the study of animal behavior as a standalone discipline—today known as *ethology*—did not yet exist. Scientists who did study animal behavior primarily came from the disciplines of zoology and psychology. Thus, two very different approaches were employed. The zoologists/naturalists (mostly Europeans) focused on observations of animals in their natural environment. The psychologists (mostly North Americans) favored laboratory experiments.[32]

The work of the psychologists led to the development of behaviorism. They considered mental processes impossible to study scientifically, and emphasized phenomena such as the response to a stimulus, as in habituation. This method eventually became well known through the phenomenon of Pavlovian conditioning, named after Ivan Pavlov.

The naturalists criticized the psychologists for focusing primarily on one species, white rats, while neglecting the study of other species. Another criticism: psychologists focused almost exclusively on learned behavior while ignoring instinctive behaviors.[33]

The scientists largely responsible for developing the naturalist approach to animal behavior were Konrad Lorenz and Niko Tinbergen. Lorenz (1903–1989), an Austrian naturalist, is considered the founder of ethology, having laid its foundations in the 1930s. His primary studies concerned instinctive behaviors in birds. In a 1936 paper he identified six principles concerning the study of instincts and ethology:

- Instinctive actions manifest themselves without the animal's having the benefit of either previous experience or learning from an older member of the species.
- Regulative control of instinctive behavior may be independent of learning and experience.
- Simple stimulus-response models are inadequate to explain instinctive behaviors.
- Instinctive behaviors sometimes occur without any stimulus.

- An animal will strive to perform its instinctive actions.
- Behavior patterns can be used in the same way as organs to determine common ancestry of species and reconstruct evolutionary trees known as phylogenies.[34]

In 1951 Tinbergen (1907–1988), a Dutch biologist and ornithologist, published one of the seminal books on ethology, *The Study of Instinct*.[35] He also wrote an influential paper on the methodology of the study of animal behavior.[36] There he outlined the "Four Questions," which he presented as the four core issues ethologists need to focus on: 1) What triggers performance of the behavior? 2) How has the behavior developed during an individual's lifetime? 3) Why does the animal perform the behavior? 4) How has the given behavior evolved over time?

These questions touch on causation, development, survival value, and evolution respectively. Scientists continue to frame the analysis of animal behavior in terms of Tinbergen's four questions—perfectly reasonable as far as it goes, but notice that the framework leaves unexamined how and why programmed (innate) behaviors arise in the first place. Simply determining how a behavior can be adaptive is not enough to even answer the *why* part. After all, there are innumerable behaviors that can be adaptive, even for a specific environment and need. Why this particular adaptive behavior? Also, charting how natural selection may have modified a given behavior over time (the evolution question) does not provide an adequate cause for the behavior's ultimate origin.

Complex Programmed Behaviors

THE IDEAS of Lorenz and Tinbergen eventually came under heavy criticism from the behavioral psychologists, with much of it finding its mark. As a result of that criticism, by 1950 ethology ceased to be the ascendant approach to studying animal behavior,[37] even though it continued to produce significant findings in animal behavior research.

The behaviorist position had far-reaching influence in the field. An important sea change in animal behavior research occurred when the psychologists who would come to be known as behaviorists abandoned the concept of strictly innate behavior.

Nevertheless, many ethologists and other animal behavior scientists still consider the general concept of innate behaviors valid. The evidence is that many animal behaviors apparently do not have to be learned. In many cases learning does occur, but as we will see in later chapters, only to refine behaviors already present. The significance of the groundbreaking work on animal behavior of Lorenz and Tinbergen was eventually recognized so that subsequently they (along with Karl von Frisch) were awarded the Nobel Prize for Physiology and Medicine in 1973.

We can refer to such behaviors as *instincts* or *innate behavior,* widely familiar terms. But we need to be aware that they are understood variously in biology, and are embattled terms.

Sixteenth-century Spanish naturalist Gomez Pereira developed a mechanistic theory of animal behavior in which he proposed that animal movement—what he called vital movements—did not involve any conscious thought.[38] Darwin's definition, previously cited, was the next significant attempt at defining instinct.

More recently, John Alcock defined instinct as "a behavior that appears in fully functional form the first time it is performed, even though the animal may have had no previous experience with the cues that elicit the behavior."[39] And James and Carol Gould define innate behavior as behavior based on inborn neural circuits responsible for "data-processing, decision-making, and orchestrating responses in the absence of previous experience."[40] This behavioral instinct is said to be encoded by genetically specified wiring. The Alcock and Gould definitions are a good starting point but are too limited in that many programmed behaviors involve a developmental component.

Early in his research Lorenz defined an instinctive behavior pattern as a "behavioral sequence based upon inherited pathways laid down in the central nervous system, such that the pattern is just as invariable as its histological foundations or any named morphological character."[41] Lorenz believed that animal behavior was a mixture of innate and acquired behaviors. Tinbergen proposed a broad definition that encompassed the species: "Each animal is endowed with a strictly limited, albeit hugely complex, behaviour machinery which is surprisingly constant throughout a species or population."[42]

But as noted, animal behavior psychologists reject any such idea. For example, a critic of the concept of instinct, psychologist Mark Blumberg, writes, "The term instinct is often merely a convenience for referring to complex, species-typical behaviors that seem to mysteriously emerge out of nowhere. But this is an illusion that is fostered by the instinct concept."[43]

The issue that many animal behavior scientists have with the word *innate* is illustrated by biologist Patrick Bateson where he compiled seven possible definitions: "present at birth; being a behavioral difference caused by a genetic difference; adapted over the course of evolution; unchanging throughout development; shared by all members of a species; present before the behavior serves any function; and not learned."[44]

Because of the problems with the definition, the term *innate* is now no longer used by animal behavior scientists. The reason, according to psychologist and animal cognition researcher Sara Shettleworth, is that "the extent to which behavior patterns or cognitive capacities are modifiable by experience varies so much as to make the terms learned and innate (or nature and nurture) obsolete."[45] Shettleworth prefers instead the terms *predisposition* or *preexisting bias*.[46] Similarly, Jerry Hogan and Johan Bolhuis propose that the term *prefunctional* is a better substitute for *innate*, which they define as "developing without the influence of functional experience."[47] All of the terms proposed by Shettleworth, Hogan, and Bolhuis appear to be appropriate.

Blumberg suggests that the concept of instinct is compromised by its association with the erroneous view of design and teleology in behavior. "Inherent in the rationalist perspective of instinct," he writes, "are the notions of purpose, goals, and design."[48] He further asserts that this is because humans have a deep-seated tendency to see teleology, such as when we witness exquisite animal behaviors that seem so perfectly designed for survival and reproduction. Blumberg further argues that this predisposition "to view the natural world and all of its wonders through the lens of design has only hindered our understanding of biological complexity, including behavior."[49] Ironically, Blumberg's critique of the concept of instinct appears to involve him saying that humans have an instinct to see in animal behavior purpose, design, and instinct where none exists. This is called having your cake and eating it, too.

All the same, it's clear that the terms *instinct* and *innate* behavior are embattled terms that are understood variously. So instead I will more often speak of *complex programmed behaviors* (CPBs). That term clearly leaves to the side simple innate behaviors as well as complex behaviors that are learned, such as a human crafting an axe or writing a letter. I will use the term instead to refer to the sorts of behavior described in the opening paragraphs—the sort of striking behaviors we observe in migrating birds and butterflies, in termites, bees, and spiders.

More specifically, let's use the term *complex programmed behavior* to describe behaviors that meet the following five criteria:

1. Complex
2. Purposeful
3. Programmed
4. Contingent
5. Heritable

Now let's briefly unpack each of those five.

Complex refers to engineering complexity, and can refer to the behavior being sophisticated, such as with animal navigation and communication. It can also refer to the complexity of the engineered system involved in producing the behavior. The term *complex* distinguishes the behaviors in question from behaviors that are simple reflex reactions.

Such complexity, notice, involves what is known as specified or functional complexity. Any random sequence of letters is complex in one sense. However, only those sequences that define something meaningful, such as the letters in this sentence, or the coding in the word-processing program I used to compose this book, are specified. The behaviors we will focus on in these pages are complex in this richer sense of specified complexity, since they are controlled by multiple (and often a large number of) genes, a form of biological information.

The behavior is *purposeful* in that it is goal-oriented or has a function, such as a spider building a web to catch food and thereby survive. Another example is bird navigation, the functions of which include enabling migration and locating food sources. A reductionist strain in contemporary biology is uncomfortable with talk of purpose or teleology, but biologist J. Scott Turner pushes back. "What if phenomena like intentionality, purpose, and design are not illusions, but are quite real—are in fact the central attributes of life?" he asks. "How can we have a coherent theory of life that tries to shunt these phenomena to the side?"[50] Indeed. Purpose is essential to complex programmed behaviors, and to life in general.

Where something is *programmed* it typically involves an algorithm and the processing of information. The mathematical definition of an algorithm is "a formal procedure for any mathematical operation, especially a set of well-defined rules for solving a problem in a finite number of steps."[51] Algorithms are a relatively recent concept, having been developed in the twentieth century. David Berlinski writes in *The Advent of the Algorithm* that its development was largely due to the work of four prominent mathematicians—Kurt Gödel, Alan Turing, Alonzo Church, and Emil Post.[52] The algorithm initially was conceived as a theoretical concept as embodied in a Turing machine, which represented an imaginary computer invented by Turing in 1936. It was later that actual computers were constructed that were capable of running algorithms through stored programs.

Interestingly, Mayr advocated the concept of programs in biology. He defined it as "coded or prearranged information that controls a process (or behavior) leading it toward a given end."[53] A common element of programming is decision making, usually needed to determine when the behavior is performed. The decision can be simple (for example—forage or don't forage), or more complex, such as the navigation route and timing in migration. Programming also applies to the process an animal goes through in developing the behavior. This includes the animal's development (ontogeny), crucial to the animal's ability to eventually perform specific behaviors.

A complex programmed behavior also must be *contingent*. In other words, is not a necessary behavior in that the given species in question might, in principle, have followed a different behavioral pattern and still thrived. Examples include the many distinct forms of programmed communication in the insect world. Specific communication methods (visual signals, auditory signals, pheromones, etc.) are contingent since different signals might in principle have been employed, and served the same function. The same is true of the various navigation methods used by animals. As we will see later, contingency poses a challenge in explaining the origin of such behaviors.

Finally, the complex programmed behavior must be *heritable*. The reason for including this criterion is to distinguish CPBs from behaviors that are primarily learned. Behaviors that develop only through learning, without a heritable basis, are not considered CPBs.

Use of the term *complex programmed behavior* in place of *instinct* or *innate behavior* won't, of course, settle all the arguments regarding the origin of such behaviors, but it may cut through some of the heat and fog surrounding the more traditional terms.

Behavior Genes?

THERE ARE three ways that complex programmed behaviors are expressed: 1) The behavior is fully functional the first time it is performed without any previous experience. 2) The behavior develops as the animal matures over time, with the development itself being programmed. 3) The behavior develops through programmed learning.

Shettleworth asserts that most behaviors can be explained by simple mechanisms: "The field of comparative cognition as it has developed in the past 30–40 years has a very strong bias in favor of 'simple' mechanisms. The burden of proof is generally on anyone wishing to explain behavior in terms of processes other than associative learning and/or species-typical perceptual and response biases."[54]

Research into the genetics of animal behavior has identified several programmed behaviors that appear to be significantly influenced by a single gene.[55] An example is the circadian rhythm of fruit flies (*Drosophila*) affected by variants of the *period* (per) gene. However, at least three other genes (*tim*, *clk*, and *cyc*) have been found to control fruit fly circadian rhythms, which are actually quite complex.[56] The neural network is also quite complex, as the clock neuron network "consists of multiple independent oscillators, each capable of orchestrating bouts of activity."[57]

A second example of a single gene that appears to govern a programmed behavior is the gene called *forager* in fly larvae. This gene has two variant alleles that influence feeding behavior. One variant ("rover") results in larvae that travel the longest distance in foraging, while the other variant ("sitter") results in larvae that remain in more closely spaced feeding packs. However, this example is also not that simple, as research has shown that the genetic switch is controlled by epigenetic factors.[58]

While apparently not controlled by a single gene, a similar example has been found in the "fight-or-flight" behavior of mice. Regulation of the level of serotonin in two types of brain neurons acts like a switch to determine which behavior to express, depending upon the threat level in the environment.[59]

While this research demonstrates that some behaviors can be greatly affected by single genes, it does not mean that even simple behaviors are controlled exclusively by only one gene. The examples cited act primarily like switches in turning behaviors on or off. Even many relatively basic programmed behaviors are guided by a complex of genes as well as epigenetic mechanisms.

Epigenetics is part of what is termed the *extended evolutionary synthesis*. It's an extension of the modern synthesis (neo-Darwinism). Neo-Darwinism, recall, is gene-centric, focusing on genetic mutations, population genetics, and statistical analysis of quantitative genetics.[60] This approach left significant gaps. As Massimo Pigliucci and Gerd Müller have noted, it was "unable to explain how organismal change is realized at the phenotype level." With the extended synthesis approach, in contrast, "gene centrism necessarily disappears in an extended account that provides for multicausal evolutionary factors."[61]

Be that as it may, the emphasis on evolution by genetic mutations has hardly disappeared, and the idea remains alive and well in attempts to explain the evolution of certain behaviors. So, for example, Sean Carroll and Patrice Showers Corneli assert that "behavior, like other phenotypic traits, varies as a function of genes and environment. Variation occurs at all demographic levels, within individuals over time, between individuals, and between populations and species."[62]

They, like most contemporary evolutionists, assume that small evolutionary variations can accumulate and translate into macroevolutionary changes to behavior. But is such an extrapolation valid? It is important to be realistic about the state of the research on this subject. Even less is understood about the origin of complex behavior than is known about the origin of complex physical characteristics. While the amazing progress that has been made in genetics has increased our understanding of the building blocks of organisms (proteins, etc.), very little is understood about the genetics of complex behavior. A statement in the 2010 textbook *Evolutionary Behavioral Ecology* remains true today: "We still know little about the rate and type of evolutionary change (punctuated equilibrium or gradualism) experienced by behavioral traits and whether it matches similar patterns in other phenotypic traits like morphology."[63] Note that this is a comment from biologists writing from within the twenty-first century evolutionary paradigm. That is, this is a confession from true believers.

The confession, however, is too limited. It isn't simply the rate and type that remain unknown. The mystery of the origin of complex programmed behaviors (CPBs) runs much deeper. To begin to see why, we turn next to particularly striking instances of CPBs—navigation systems and migration.

2. NAVIGATION AND MIGRATION

The pilot is made by precepts which tell him thus and so to turn the tiller, set his sails, make use of fair wind, tack, make the best of shifting and variable breezes—all in the proper manner.[1]

— Lucius Annaeus Seneca

THERE ARE FEATS OF MIGRATION IN THE ANIMAL KINGDOM SO AWE-inspiring and mysterious that researchers have yet to unravel all their secrets, much less explain how those abilities evolved in the first place. The navigational capacities involved appear to be innate—all the more striking because humans apparently do not have any such innate ability. Several studies have demonstrated that humans do not have an innate sense of direction. The fact that it took long ages for humans to develop accurate and reliable methods for navigating long distances indicates how difficult a challenge it is. This provides a point of reference for the remarkable examples of animal navigation, which exceeded human navigation methods until relatively recently.

Animals, it turns out, make use of some of the same methods for navigating that humans later invented. James and Carol Gould write, "We must depend on luck as much as talent, trying with clumsy approximations to replicate the compass sense that animals use innately to pinpoint our position on the globe without the seemingly magical combination of sensory abilities and inborn processing circuits that for other species comes as standard equipment."[2]

Having worked on various aircraft navigation systems throughout my engineering career, when I first began to study complex programmed behaviors (CPBs), dramatic feats of animal navigation were the first such behaviors to catch my attention. My familiarity with the challenges of designing navigation systems made me curious about remarkable feats of animal navigation. And when I delved into the science of it, I was surprised to learn that in many cases animal navigation is, even today, as good as or better than human-engineered systems.

Arctic terns complete the longest migration of any animal. In the Northern Hemisphere's spring they breed in the Arctic, not far from the North Pole. Several months later they fly to the Antarctic to spend the summer in the Southern Hemisphere. Thus, they complete a round trip of 25,000 to 30,000 miles. Based on banding information, the oldest known Arctic tern was over thirty-four years old, and may have accumulated over a million miles in its lifetime.[3] Then, too, as noted above, monarch butterflies migrate as much as three thousand miles between Canada and Mexico, and it takes multiple generations of butterflies to make the trip. How do organisms like the Arctic tern and the monarch butterfly manage to achieve such amazing feats of navigation? How could these navigation capabilities have evolved through a blind and gradual evolutionary process?

Figure 2.1. An Arctic Tern

In addressing that question, the first thing is to recognize that animals, human and otherwise, use a variety of navigational techniques, including many sources of positional information—e.g., landmarks, the sun, polarized light from the sun, magnetic compass, stars, chemicals, smells, temperature, and gravity. More broadly, James and Carol Gould identify six distinct navigational strategies:

1. Taxis: Movement following a cue, such as light (phototaxis) or chemical (chemotaxis).

2. Landmark navigation: Using familiar landmarks to navigate. Also referred to as piloting.

3. Compass orientation: Following a constant bearing relative to a cue.

4. Vector navigation: Following a sequence of compass bearings, usually independent of landmarks.

5. Dead reckoning: Calculating current position by estimating the distance and direction traveled at each leg, without reference to landmarks.

6. True navigation: Navigating based on knowledge of the destination. This requires some form of a map sense.[4]

Figure 2.2 depicts a generic animal navigation system. The basic elements include a sensor that determines the current position, sensor information processing, information defining the desired path, and a control algorithm. Following is more about some of these types of navigation along with examples of how particular animals employ them.

Landmark Navigation

LANDMARK NAVIGATION employs prominent visual cues. The use of landmarks is the simplest type of human navigation. This is most readily used where landmarks can be distinguished at a distance. It would be difficult, for instance, to use it in a dense rainforest where no distant objects can be seen. There are instances where it is used at sea. Ship captains and some birds use landmarks over water, when sufficiently close to

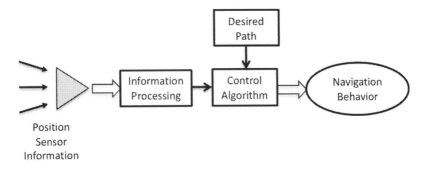

Figure 2.2. Generic Animal Navigation System

land to maintain visual contact. Ships also use lighthouses as key points of reference.

Most animals employ landmark navigation in one form or another, even though they also use more sophisticated navigational techniques. Many of these animals can form a type of cognitive local area map, based on the landmarks they observe over time. Niko Tinbergen conducted several experiments with bees and wasps to determine how they made use of landmarks. Digger wasps keep their prey in burrows, which they need to be able to return to reliably. They make use of local landmarks to keep track of the burrow locations. Tinbergen's experiments found that the wasps had a strong preference for dark-colored three-dimensional markers.[5] One reason why the landmarks need to have prominent features for the wasps to identify is that their vision is extremely poor (20/2000).[6] The Goulds comment, "There is no doubt that bees and wasps are born already knowing something about how to choose and use landmarks."[7] Therefore, they are born with some form of procedural knowledge, which is likely programmed. Experiments have also demonstrated that ants use landmarks when navigating between their nests and feeding locations.[8]

Dead Reckoning

Dead reckoning is a shortened version of the original term "deduced reckoning." It does without landmarks and depends on taking accurate measurements of the ship's movements—its direction but also, very significantly, its speed.[9]

Computing speed is based on dividing the distance traveled by the elapsed time. That may sound easy, but until the last few hundred years there was not an accurate mechanism available for measuring either distance or time. Accurate clocks did not yet exist. Estimating speed typically consisted of throwing an object overboard at the bow and waiting for it to reach the stern. Speed was then computed by dividing the length of the ship by the time it took for the object to travel this known distance, where time was often measured using a sand hourglass. Another method used a line that was knotted at fixed intervals and dropped into the water at the stern. The faster the ship was moving, the farther the current would pull the point that the rope entered the water away from the ship, drawing more rope, and rope knots, above the water line. That is the basis for the term "knot" as the unit of speed, where one knot equals one nautical mile (1852 meters) per hour. While this method is no longer widely used, knots remain the basic unit of speed in both ship and aircraft navigation.

With the advent of aviation early in the twentieth century, one might think that technology would have immediately rendered navigation relatively easy. However, that was not the case. Early aviation navigation was so primitive that one way that aircraft position fixes were obtained was by painting signs on the tops of buildings such as barns and constructing other types of ground monuments. These had to be seen visually, so they were of no use at night or in foggy or cloudy conditions. Eventually ground-based radio navigation systems were developed that enabled aircraft to navigate reasonably well over long distances even in heavy fog and at night.

Oceanic flights still had a problem with navigation since ground-based radio systems as well as radar are not useable beyond several hundred miles. Therefore, celestial navigation was commonly employed, a method taken off the board on cloudy days.

Long-distance aircraft navigation improved significantly with the development of inertial navigation systems, which are based on the use of gyroscopes and accelerometers.[10] These sensors enable the system to compute the speed and direction of movement. The accuracy of aircraft inertial navigation systems still has limitations. Accuracy is expressed in terms of error over one hour of operation. High quality aircraft systems have an error rate of roughly 0.5 miles per hour of flight.

Some animals use a form of dead reckoning known as *path integration*. In this method animals keep track of the compass heading and distance traveled. How does an animal keep track of the compass heading? Obviously they don't purchase compasses at their local sporting goods store and cart them around their migratory treks. The answer is that long before humans "invented" the compass, these animals already employed ingenious built-in compasses (more on this below). This should give us pause, as should the fact that these animals also have a built-in capacity to read their compasses.

By continually keeping track of the compass heading information and distance traveled, such animals can compute a direct path back to their starting points, even when they take a highly circuitous foraging path away from their home nest. They do not simply retrace the longer outbound path to return. Figure 2.3 illustrates an example of an animal's

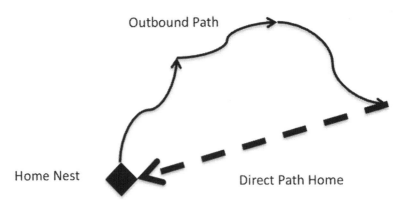

Figure 2.3. Animal Path Integration (Dead Reckoning)

movement in such a case. Experiments have shown they are able follow the direct path without the use of any landmarks.[11]

In discussing the ability of honey bees to perform such feats, the Goulds suggest that this shows that "honey bees can do the necessary trigonometry for this dead reckoning with considerable precision."[12] This means they can compute both the direction and distance components after circuitous flights through unfamiliar territory.

There are two limitations in the use of dead reckoning for airborne animals. One is the effects of wind on the path as any significant wind will result in an error of the computed change in position.[13] The other is that an accurate calculation of ground speed and estimated location requires knowing the altitude at which the animal is flying. One group of researchers developed a proposed neural model for how path integration could be implemented in bees.[14] The model includes how the mathematical computations could be accomplished based on the navigation sensor information. They also built and tested an experimental robot to simulate the model. The model includes the use of trigonometry and other complex math.[15]

Compass Navigation

COMPASS NAVIGATION, as the name implies, consists of following a compass heading—that is, of following some device or cue that indicates geographic direction. (This is known as vector navigation if a sequence of compass bearings are followed.) In compass navigation humans follow hand-held devices that register the earth's magnetism to indicate north, while animals employ several types of cues that indicate geographic direction, including magnetism, the sun, stars, and polarized light.

Magnetic Compass

The Chinese were the first to make use of magnetism in a compass, in approximately 1088.[16] It was eventually introduced to European sailors. During its early use, it was not known that there was a difference between true north and magnetic north. After this discrepancy was discovered, the Portuguese developed tables that tabulated the differences relative to specific locations. However, in the 1600s it was discovered that the location of the magnetic poles also varies continuously.[17] The location of the north magnetic pole moves approximately 10–15 miles per year, which obviously creates problems for reliable navigation.[18]

It has now been established that many animals have a magnetic sense that can be used as a compass for determining direction. This exists in widely diverse groups of animals, including all major groups of vertebrate animals, some mollusks, crustaceans, and insects.[19] The exact mechanisms of this magnetic sense are still poorly understood.

We know that birds possess a magnetic sense they use on their migration routes.[20] While we do not understand all of the details yet, it is clear that birds have a well-engineered ability to determine navigation position based on Earth's magnetic field. The value of the magnetic compass is that it is always available, regardless of visual conditions. Its presence has been demonstrated in numerous migrating songbirds as well as homing pigeons. Apparently juvenile birds rely primarily on the magnetic compass in their initial migration, as they can initiate a flight in the correct migratory direction without any other clues.[21] This suggests that the basic migratory information comes pre-programmed.

Much of the initial research focused on characterizing the nature of the magnetic compass, and its possible location in the given animal under investigation. There is experimental research suggesting that birds have magnetic detection receptors in the retina, nose, beak, and inner ear.[22] There are two primary candidates for the magnetic sensor—magnetite and cryptochrome.

Magnetite is a highly magnetic iron oxide mineral (Fe_3O_4) also known as lodestone. Invertebrate animals that have been found to have magnetite embedded in their bodies include monarch butterflies, spiny lobsters, moths, and termites.[23] Magnetite has also been found in several vertebrates including birds, salmon, dolphins, sea turtles, mole rats, and even newts.[24] James Gould indicates that magnetite-based sensors are the only plausible candidates with the necessary sensitivity to detect the magnetic field and extract relevant information.[25] An aspect not understood is how the magnetite crystals are manufactured or integrated into organisms.[26]

The other leading candidate for the magnetic sensor, cryptochrome, is a protein found in many animals and plants. It functions as a photo-receptor to blue light and, in the animals that possess it, is found in the cones of their eye retinas. The theory is that it can function as a mag-netic detector through a complex process where light detection causes the formation of a pair of electrons that then are affected by the mag-netic field.[27] The best evidence from recent research for this method of magnetic detection is in birds.[28]

Recent research suggests how the magnetic detection information might be sent to the brain. An experiment identified neuronal response in the brains of pigeons and confirmed that pigeons can detect the three characteristics of magnetic fields—direction, intensity, and polarity.[29] The neurons were found to respond to the magnetic field vector, which includes intensity, azimuth, inclination angle (elevation), and polarity. This is significant because this information can be used to determine global position. However, there is still much to be learned about how magnetic fields are detected and how exactly the information is pro-cessed in the brain.

This information still needs to be applied to determine the desired navigation path. It could be based on either the sun compass or magnetic compass, as shown in Figure 2.4. Even this is no small task. Keep in mind that a compass only indicates the direction of north. Determin-ing the path heading involves computing the angle between north and the destination. This appears to be the method used by most migrating animals.

Sun Compass
Another source of navigation information that many animals use is the sun compass, which involves a rather complex method. The need for complexity comes from the fact that the sun is continuously moving from east to west throughout the day. To use the sun to determine direction

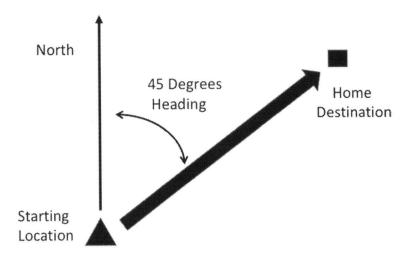

Figure 2.4. Navigation Path Vector Determination

requires knowledge of its relative position to true north. This is no trivial matter since the relative position of the sun to true north varies not only throughout the day but also varies with latitude and changes throughout the course of the year due to the tilt of the earth's axis. The time of day when dawn occurs, as well as the sun's maximum elevation, changes each day for a fixed position on earth. The only thing that is constant is that the maximum elevation occurs halfway between dawn and dusk.

As an example of how all this must be taken into account, consider the southward migration of monarch butterflies. As Gould and Gould explain, these butterflies must "steer right of the sun at 9 AM, toward the sun at noon, well to the left of the sun at 3 PM, and so on, always heading due south."[30]

Theoretically the elevation of the sun could be used to adjust for time, provided the animal had an inbuilt sense of its latitude and time of year, since the elevation varies with latitude and season. However, experiments have shown that animals do not use elevation, but instead rely primarily on determining the angle between due north and the sun's vector on the horizontal plane, an angle known as the azimuth angle.[31] Since the azimuth angle of the sun moves continuously throughout the day, animals keep track based on time. This requires an internal clock to keep time and compute the compensation. To make it even more complicated, the sun's rate of change in azimuth angle also varies with latitude and season. Therefore, the time compensation is not fixed but must continuously be adjusted to account for these differences. And yet animals including monarch butterflies, bees, and many birds are equipped to perform the necessary computations with sufficient accuracy.

The Goulds describe one experiment with bees that indicated bees use the average position of the sun over the previous forty minutes, which improves the overall accuracy of the position estimate, rather than using a single snapshot.[32] Experiments have demonstrated that even when animals are blocked from viewing the sun for hours, they can still correctly compensate for its change in position.[33] Other research has shown that ants and honey bees have an innate program that estimates the position of the sun throughout the day. The algorithm is based on an approximation that the sun maintains a constant azimuth in the morning, changes relatively quickly by 180 degrees at midday, and then again stays constant in the afternoon. However, this approximation is not fixed, as the insects are able to learn a more accurate position estimate based on the information at the local latitude.[34]

Time compensation has been demonstrated in several experiments where the animal was artificially displaced, effectively changing the location of the sun as viewed by the animal. The animals responded by moving in a direction based on where the sun would have been prior to the displacement. The experiments demonstrate that the animals base their navigation on an algorithm that estimates the sun's relative position throughout the day.

Polarized Light Compass

In addition to direct observation of the position of the sun, some animals can use polarized light to determine the sun's position, even when heavy cloud cover makes it impossible to tell where the sun is by the ordinary means of glancing up in the sky and spotting the sun. As light from the sun enters the atmosphere it is unpolarized, meaning the wave pattern of a given stream of photons is not aligned along a single plane, and is therefore random. Following transmission through the atmosphere the photons become partially polarized perpendicular to the travel path. Figure 2.5 illustrates the effect of polarization.

Polarization (indicated by the arrows) is always perpendicular to the plane containing the observer, the sun, and the sky location being observed. Along the solar meridian the polarization is parallel to the horizon. In other directions, such as the circle (shown as an ellipse) about midway between the top and bottom, the polarization direction varies through all angles. Interestingly, the polarization is not unique for every point in space because at each elevation every polarization direction occurs twice. This presents an ambiguity in determining the sun position and using it as a navigational cue.[35] It appears most animals somehow

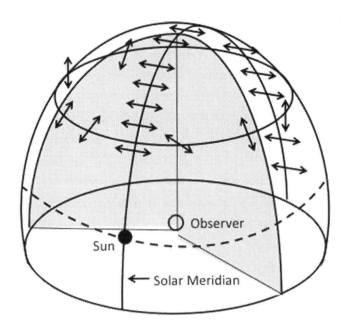

Figure 2.5. Sunlight Polarization

can resolve the ambiguity as long as they can view multiple regions of the sky.

Some animals can detect the orientation of this plane of polarized light and from this infer the position of the sun, even on heavily cloudy days. However, simply detecting the polarization orientation does not completely solve the problem. In addition, the polarization vectors change as the sun moves throughout the day, thus further adding to the complexity of the compass algorithm. The polarization compass also has the same challenge as the sun compass in compensating for the path of the sun throughout the day (dashed line in Figure 2.5).

A clever laboratory experiment demonstrated that honey bees can navigate solely on detection of the polarized light vector.[36] They, as well as ants, have specialized structures in the eye and brain that function as polarization detectors.[37] The discoverer of the polarization compass in bees, Karl Von Frisch, demonstrated that bees need to view only about a ten degree patch of clear sky to determine the polarization and thus orient their dances accurately (more on that later).[38]

Using polarization information to determine the position of the sun is no simple matter. A computer program that could manage the task would require some sophisticated software code and mathematical calculations. Bees seem to possess something like that. As the Goulds put it, "Bees, in fact, have added a processing trick that allows them to guess the sun's azimuth."[39] It requires complicated geometry that one might imagine is well beyond the capacity of tiny-brained bees and ants. But apparently bees and ants can manage the necessary computations since it's been shown that they still know the sun's position even when it is not directly visible.

A different use of polarized light has been discovered in birds. One study of migrating songbirds concluded that the magnetic compass is recalibrated with regard to polarized light cues at sunrise and sunset. More precisely, when there is a conflict between magnetic and polarized light cues at either time, the magnetic compass is recalibrated. This recalibration occurs both before and during migration, and in both juvenile and adult birds.[40] Further research has indicated that birds tend to ignore sunlight polarization cues when the sun is near zenith. The apparent reason is that the e-vector of sunlight at zenith aligns east-west, and is thus perpendicular to the magnetic compass, causing confusion as the information is perceived by birds.[41]

Their method of recalibration is near optimum because maximum polarization occurs at sunrise and sunset, and averaging the two measurements minimizes error. And ignoring it near zenith (noon) also eliminates errors. The use of some form of spherical geometry and method of calibration are examples of the sophisticated engineering design of this navigation behavior. Spherical geometry is complicated by the fact that on a sphere there are no straight lines, so standard (Euclidean) geometry does not work. Human mathematicians perform the calculations using complex spherical trigonometry. The precise specifics of how an animal's (tiny) brain performs such computations remain unknown, but again, it appears to involve innate programming.

Celestial Navigation Compass
Another form of celestial navigation is based on the stars. There are two elements to this form of navigation. One is use of the pole point, which is the spot in the sky which the sun, moon, and stars appear to rotate around, as the earth rotates. The star presently closest to the pole point is the aptly named Pole Star (Polaris), located 0.7 degrees away from the pole point axis. However, because it is not located exactly on the pole axis (also, the rotation pole moves around, or "wobbles," over the centuries), it presents problems in enabling reliably accurate navigation. The second element of celestial navigation is the use of star constellations. Experiments have shown that birds can determine the pole point by observing the constellations and the Pole Star rotating around it.[42] Apparently, it is this stationary point that birds use as the reference, not any one star near the point. They also can update the information because the positions of constellations change throughout the year. Similarly, in the Southern

Hemisphere the same phenomenon occurs, not with a single star, but with the Southern Cross constellation rotating about the pole point.

As with other compasses, the origin of the use of pole points for animal navigation is a mystery. The behavior is programmed and complex. Humans have the cognitive capacity to study and master how these references can be applied to navigation using mathematics. However, as noted, animal types described above do not learn it. They just know it. The ability is innate.

Map Sense and True Navigation

ANIMALS ARE also capable of what is known as *true navigation* if, after being displaced to a spot where they've never been, they are able to determine their position relative to a destination goal without having to rely on familiar surroundings.[43] This ability is especially impressive.

Compass navigation or inertial navigation (path integration) can only provide guidance along a heading or to and from a fixed point; but in true navigation, the map sense provides the capability to navigate from any location to a specific destination (as we can do with GPS). Also, the distance these animals are displaced has no impact on their accuracy in finding home. Simple compass-and-vector navigation does not provide that level of accuracy, because accuracy degrades with distance, as described previously regarding aircraft inertial navigation systems.

The type of map we are most familiar with is based on defining positions on the earth in terms of latitude and longitude, and is therefore called bicoordinate. With such a system, a course (heading) can be determined between any two points, along with the distance. With the widespread use of GPS, most people are now familiar with such maps. The real breakthroughs in worldwide navigation occurred when the ability to determine latitude, longitude, and magnetic heading developed. The concept of using imaginary lines of latitude (north-south position) and longitude (east-west position) to define locations on the earth was first developed by Eratosthenes (276–196 BC). However, it wasn't applied until Ptolemy used it in his world atlas in 150 AD. Latitude was apparently first used for navigation by Portuguese sailors in the early 1400s.[44] They determined it by using the Pole Star as a reference.

Determining longitude was an even more difficult problem since it requires the use of an accurate clock. One can convert time into longitude since the earth completes a full rotation of 360 degrees in twenty-four hours; therefore one hour equals fifteen degrees. Local time can be derived based on the determination of the local noon (when the sun reaches its highest point in the sky). Converting longitude into distance is complicated by the fact that degrees of longitude have decreased separation distance as you move from the equator to the poles. Even after relatively accurate clocks were developed they were nearly useless on ships since these clocks used pendulums, and ships are in constant motion at sea. The harsh environment also contributed to the difficulty of maintaining reliable devices. The impact of the lack of accurate knowledge of longitude was indicated by the untold number of ships and seamen lost at sea due to poor navigation. The most famous such incident occurred in 1707 when four British warships ran aground and sank, killing about two thousand sailors and soldiers.[45]

It was such disasters that motivated the British Navy to establish a reward for someone who could develop a clock useable for reliably determining longitude. In the book *Longitude*, Dava Sobel tells the fascinating story of how such an invention was accomplished by John Harrison in 1759. Harrison's clock is on display in the Greenwich observatory museum, where I have observed it and admired the incredible engineering that it embodies. The first few models were about the size of a large microwave oven. It took several iterations to reduce the clock to a size that was practical for ships. The goal that was established by the British Navy was to be able to determine longitude to within a half degree, or approximately thirty nautical miles.[46] While this doesn't sound that impressive, it far exceeded the accuracy of previous methods.[47]

We've known for some time that birds possess some form of map sense. Pigeons are the best example, as they can navigate back to their home from any location. It also appears that many long-distance migratory birds also possess a map sense. The evidence supporting this is that these birds can fly thousands of miles and are still able to locate their destination with great precision. Examples include the bar-tailed godwit and bristle-thighed curlew, both of which migrate between Alaska and islands in the South Pacific.[48] Another indication of a map sense is that long distance migrants fly individually at night, each following its own course, and thus not simply following others in a group.

It was originally thought that only certain vertebrates had a map sense and the capability for true navigation. However, more research has also found a map sense in various invertebrates, including lobsters and mollusks.

Research with loggerhead sea turtles and spiny lobsters determined that they apparently have a map sense based on Earth's magnetic field, elements of which vary geographically, including the total field intensity, vertical intensity, and horizontal intensity.[49] Recent research has found that spiny lobsters can orient in the proper direction toward home from as far away as twenty-three miles and can detect longitude as well as latitude.

Some animals can detect the magnetic field intensity. Some also can detect the magnetic inclination angle. This can be a source of latitude information as the inclination angle varies from the North Pole (90 degrees) where the field points straight down, to the equator (zero degrees) where the field is horizontal, to the South Pole (-90 degrees) where the field points straight up. There is evidence that magnetic field detection in birds is not based on the north-south polarity of the field (as was previously assumed), but rather depends on vertical inclination.[50]

There is no regular variation in the Earth's magnetic characteristics with longitude, which means there is no simple relationship between the detected field and the actual longitude. Despite this, research found a mechanism in loggerhead sea turtles that does allow them to use the magnetic field to calculate longitude,[51] information the turtles use to navigate on their journey in the North Atlantic. The researchers theorize that longitude, and therefore a bicoordinate map, are made possible because the regions traversed by the turtles are characterized by magnetic fields that have unique combinations of intensity and inclination, both of which the loggerheads detect. If some animals are able to form a true bicoordinate map, it would explain how some are able to navigate accurately over extremely long distances. The Goulds comment, "As a global-scale strategy that can provide (indirectly) both latitude and longitude, the magnetic GPS is far and away the most satisfactory explanation."[52]

Despite the evidence that ants and bees can determine direct routes via path integration, it appears they do not possess a true map sense.[53] Recent research by James Cheeseman et. al indicates that bees do possess a form of map sense different from a true map. When one compares the relative distances traveled by ants and bees to those traveled by birds and ocean dwellers such as sea turtles, a global map sense (equivalent to GPS) would not seem to be necessary. Instead, their map sense may be based on a metric cognitive map, defined as being able to locate oneself in space using directions and distances.[54] That means the animal can navigate from point A to point B because it can determine direction and distance. The research showed that bees can navigate using local landmarks after being both physically displaced and placed in a simulated altered time of day, which normally disrupts their sun-compass navigation. Similarly, Randolf Menzel and Uwe Greggers also concluded that honey bees possess a form of spatial cognitive map that is based on integrating a compass bearing map and local landmarks.[55] What is intriguing is that this type of map had only been previously demonstrated in mammals and birds, which have significantly greater cognitive capability.

While the possession of a true map sense in animals is an amazing discovery, the sophisticated use of spatial cognitive maps is equally if not more impressive. To start with, just because you know where you are and what your destination is on a map does not mean you automatically know how to go from A to B. It is not a trivial exercise even for most humans to compute the path. First one needs to determine the relative path angle from A to B. Then this needs to be converted into a track based on the navigation cue (again refer to Figure 2.4).

There is another level of complexity employed by some long-distance migrants, particularly birds, an extra level required because the Earth is a spheroid and so straight-line tracks (as seen on flat map projections) are not the shortest or most efficient when traveling long distances. The shortest routes are called "great circle routes," because when projected onto a flat map the route is curved. These are the routes modern aircraft generally follow when flying long distances. The easiest way to visualize the difference is to think of taking a flight from New York to Tokyo. If you look at a flat map it would appear that the shortest route is to fly in a straight line on the flat map, almost due west and ever so slightly south, since Tokyo is close to the latitude of New York but a bit further south. However, if you look at a globe it is readily apparent that the shortest route goes significantly north over Canada and Alaska. This is a great circle route. The problem is that computing such a route involves the use of spherical trigonometry. In addition to the complex geometry, the traveler's trajectory must be continuously adjusted to stay on the route. It is only with the advent of computers that aircraft and ships have been able to closely follow great circle tracks. Yet somehow birds have a programmed algorithm that can compute these tracks. Several hypotheses have been proposed to explain how birds follow great circle routes.[56] While there may be ways of minimizing the mathematical computations, they are still complex algorithms. The origin of such algorithms and how they are implemented in the brains of birds needs an explanation. This presents a challenge for evolutionary explanations, which we will take up later.

Long-Distance Homing and Migration

ANIMAL MIGRATIONS involve more than moving from point A to point B. To succeed, a migration must occur at the right time, under the right conditions, and follow a correct route to the destination. As Francisco Pulido notes in the journal *Bioscience*, migratory birds possess "an innate program that 'tells' them how fast to develop, when to leave the breeding area, how fast and in which direction to fly, and when to stop migrating."[57] Many birds also possess an innate homing ability to navigate to their home location.

The showiest instances of animal navigation involve migration over long distances, but whether it's the derring-do of an Arctic tern, or the modest migrations of zooplankton in the ocean that migrate up and down vertically on a daily cycle,[58] animal migrations tend to share five common characteristics.[59] The first is persistent movement, meaning they move to a different habitat. The second is that the animal follows a consistent course to the destination. Third is that for long periods during migration animals are undistracted by stimuli that would normally cause them to halt their movement. For example, migrating animals often go long stretches without feeding, despite losing weight in the process. This suggests that animals are programmed to prioritize their migration goal over other physical needs. The fourth characteristic is the distinct behaviors associated with migration during the departure and arrival. For example, before departure most migrant birds go on eating binges, storing up fat reserves for the journey. This behavior leads into the fifth characteristic: they reallocate energy to support the upcoming trek. This is particularly true for birds that make long migratory flights involving few or no stops.

Some birds such as geese and storks learn from other experienced birds of the same species. However, many juvenile birds can fly the correct migration route without the benefit of any learning from adults.

There are only two plausible ways that the migratory ability of juvenile birds is programmed: (1) they are programmed to fly in a specific direction (based on a compass heading), or (2) they are programmed to reach a specific destination.[60]

Humans have been aware for thousands of years that birds can navigate over long distances. Over four thousand years ago Egyptians domesticated and bred wild rock doves, the ancestors of modern homing pigeons, to transport messages.[61] Homing pigeons, like all migratory birds, apparently employ several navigation strategies. Typically, they use a simple cue while they are young and inexperienced, and then as mature adults switch to more complex cues and sophisticated strategies.

When returning to their home loft, juvenile pigeons use a combination of dead reckoning based on inertial movements during the outbound part of their journey, and a magnetic compass. After pigeons mature and have more experience they also use the sun as a compass cue. Pigeons fly only during the daytime, eventually relying on the sun compass as the primary form of navigation. However, on cloudy days they can switch back to the magnetic compass to navigate.[62] When they get close to their home they switch to landmark navigation. They can also use olfactory cues in close vicinity of the home loft. Pigeon strategy, then, is to use the navigation cue or cues that are the most accurate available in a given circumstance.[63] This often means using more than one method at a time to improve accuracy.

In sum, pigeons' ability to navigate improves with experience, so there is a learning process to enable more accurate navigation, but most aspects of their behavior are innate and therefore programmed, including the learning process itself. Pigeons have been bred to improve their navigation ability, but as Gould and Gould comment, "Most researchers in the field see no fundamental difference between the map and compass abilities of [artificially bred] pigeons and those of conventional migrants."[64]

In addition to the Arctic tern, there are numerous examples of wild bird species that engage in amazing feats of long-distance navigation. Manx shearwaters are shorebirds that migrate between the east coast of South America and several small islands near Wales and Scotland, where they breed and raise their young.[65] Besides the long distance, there are several other notable aspects of shearwater migration. One is that the adults initiate their return flight to South America before the juvenile birds depart, so the young birds must make the journey without any guidance from the adults. This is not uncommon in bird migration, and it means that the information defining the navigation route must be pre-programmed.

Another notable aspect is the difference between the two routes, as shearwaters do not simply retrace their route on the return flight. The northbound route goes through the western part of the Atlantic Ocean. The southbound flight follows the coast of western Europe and northern Africa. The routes and their seasonal timing appear to be correlated with environmental factors such as optimal foraging opportunities.[66]

Most notable about Manx shearwaters is their navigation system performance. Research indicates they may possess a true map sense. That is based on findings that show their navigation performance is much better than could be achieved with simple path integration.[67] The measured error was found to be only 0.006 degrees per kilometer. That translates to an error of only fifty meters over a five-hundred-kilometer distance. That is an astonishing ten times more accurate than a commercial aircraft inertial navigation system! The source of the shearwater map sense is still a mystery, but this same research also indicates it includes the ability to compute the distance to home.

Another migratory shorebird that breeds in the Arctic regions is the bar-tailed godwit. There are four sub-species that migrate to different regions, with the most interesting being the Alaskan godwit. It migrates between either New Zealand or Australia in the Southern Hemisphere, to western Alaska in the Northern Hemisphere. The migratory route consists of a series of nonstop flights. One bar-tailed godwit that was tagged with a satellite tracking device was recorded flying nonstop for 6,350 miles from New Zealand to China. Then, after feeding for five weeks, it flew another 4,013 miles to the Alaska Peninsula.[68] Even more impressively, the same bird flew the return route to New Zealand, covering 7,189 miles nonstop in eight days. That is an average speed of about thirty-five miles per hour for close to two hundred hours!

In the case of many migratory birds, their migrations deliver them from harsh winters. But in the case of some migratory birds, the pattern is less simple, and the logic of it less easily explained. White-rumped sandpipers are 1.5 ounce shorebirds that migrate between the Arctic and Argentina and Chile, with the north-bound flight being nonstop.

Brian Harrington describes several complications and challenges facing long-distance migrants such as the sandpipers, including "flying different routes northward and southward, utilizing very different habitats during the course of a year, and feeding on different prey species."[69]

For Harrington this presents a mystery. "I wonder what advantage is gained by traveling to Argentina or Chile instead of Texas, Florida or Georgia?" he asks.[70] He goes on to comment that from a human perspective, it seems incredible, even bizarre, that such small birds would undertake these amazing migrations, and that their lifestyle includes such high risk and frequent deaths.[71] However, analysis of the migration route and the availability of food resources appears to provide at least a motive for the distance and routes covered by the white-rumped sandpipers. Harrington explains that the stepping-stone pattern appears to be a highly refined itinerary which "takes them—by a series of long-distance nonstop flights—between predictable, but seasonally ephemeral, food resources scattered at distant points" of the globe.[72]

Kenneth Able writes that despite nearly fifty years of intensive study that yielded numerous startling discoveries, "we still cannot explain in a detailed, mechanistic way how birds do what we know with certainty that they do; return with incredible precision of timing to pinpoint locations on the earth after traveling thousands of miles in the interim."[73] Although scientists are discovering more about navigation sensors, many aspects of bird navigation ability and design remain a mystery.

3. Navigational Genius—
Not Just for the Birds

> The minute dimensions of some animals' brains are as astounding as the homing capacity of some of their owners.[1]
>
> — Bernd Heinrich

Till now we have mainly focused on the navigational feats of various birds, but nature's master navigators can also be found under water and among various insects. Let's start small and work our way up to loggerhead sea turtles and then tie it all together by comparing the most impressive instances of animal navigation with the most advanced navigational systems on modern aircraft.

Approximately 3.5 trillion insects migrate annually just within the United Kingdom.[2] The magnitude and scope of insect migration is impressive. So too is their steadiness of purpose in these long treks. It used to be thought that insect migration basically consisted of moving with the wind, in whatever direction it happened to be blowing, but research has shown otherwise. Studies of insect migration in the UK found that the direction of migration is consistently north in the spring and south in the fall, and the insects frequently fly in a direction different from the prevailing winds.[3] Clearly these insect navigators have a purpose they are pursuing, wind direction be hanged.

Bee Navigation

Purpose and tenacity are common hallmarks of insect navigators. Take honey bees. They face many navigational difficulties related to foraging and establishing new colonies. In covering as much as 150 square miles around a nest,[4] they use several methods of navigating, including visual landmarks, sun compass, and polarized light compass. Each is employed depending upon the circumstances. Under good visual conditions with sufficient references the bees navigate primarily by visual landmarks, while also maintaining the sun-compass information. On cloudy days when the sun is not directly visible they can use the polarized sunlight compass.

When a scout bee locates a good feeding source, it navigates back to the hive and communicates the location of the feeding source through what is known as a *waggle dance*. The Goulds call this curious dance "the second most information-rich exchange in the animal world,"[5] second only to human language. That is quite a statement considering the communication is by insects with only 950,000 neurons, compared to humans with about eighty-five billion. Honey bee brains are less than one cubic millimeter in size.[6] That is, a thousand of their brains together wouldn't amount to even a single cubic centimeter. A curiosity is that honey bees have brains only about half the volume of bumble bee brains, yet exhibit a larger repertoire and more complex behaviors than bumble bees.[7]

While the details of the waggle dance are still not completely understood, a significant amount of research, starting with Karl von Frisch, has revealed the basic methodology. The behavior develops in adult honey bees who have emerged from the pupa stage and chewed through the protective cell to join the colony. Honey bees are able to interpret the dance after about one week. The development includes electrophysiological changes in brain neurons, evident when comparing mature foragers with newly emerged bees.[8] Therefore, the behavior appears to be a combination of innate capabilities and pre-programmed learning.

The waggle dance consists of several elements that convey navigation information. The waggle itself is where the bee shakes its body at a rate of about fifteen times per second. This occurs during the middle of a sequence where the bee movement is in the form of a figure eight. The orientation of the bee during the waggle portion of the dance conveys the direction of the food source. The way this is done is not straightforward. The orientation of the movement is relative to the vertical direction on the honeycomb wall in the hive. The angle between the direction of the dance and vertical represents the angle of the vector of the food source relative to the sun. If the direction of the food source is toward the sun, the dance orientation is vertical (regardless of what direction the sun actually is). If the direction is forty degrees clockwise from the sun, the dance orientation is forty degrees clockwise (or right) of vertical.

The dance and associated vector information are relative to gravity and not to Earth's geomagnetic field, as some had speculated.[9] However, the vector angle communicated in the dance is relative to the sun. This means that honey bees have a mechanism for detecting the gravitational field, although the exact mechanism is yet to be determined.[10]

The duration of the dance conveys the distance of the source, where one waggle run (in the figure eight) signifies a standard distance, which varies between five and fifty yards, depending upon the species. How the bee calculates distance is still to be determined. Some suggest it is based on optic flow, the progression of objects across the animal's visual scene.[11]

In any case, the waggle dance communicates the full vector information (direction and distance) necessary for other bees to locate the food source. Another impressive aspect of the waggle-dance communication: it compensates for the movement of the sun over time. Thus, when the bees perform the dance and convey the vector angle leading to the food source, they adjust the angle based on the time of day.

Everything about this behavior is complex. It starts with bee foragers being able to determine the distance and compass heading relative to the food source. The bees must then translate this information into a message they convey to other bees via the dance. Other bees in the nest then must be able to interpret this information and use it to navigate to the food source. How can their tiny bee brains manage all this? Australian biologists Andrew Barron and Jenny Plath note that despite bee researchers investigating the subject at great length, "We still know very little about the neurobiological mechanisms supporting how dances are produced and interpreted."[12]

The temperate-zone honey bee waggle dance, it's been postulated, may have evolved from the dwarf honey bee (*Apis andreniformis* and *Apis florea*), which resides in the tropics. Dwarf bees build their hives in the open with a nearly flat top. Their waggle dance is oriented directly toward the food source.[13] Several decades ago Martin Lindauer proposed a similar process, where the complexity of the waggle-dance communication evolved incrementally.[14] While these theories may appear reasonable at first blush, given the complexity of the behavior it is unclear how a Darwinian process can be a plausible explanation. There is a suite of individual capabilities and behaviors involved (including navigation, data processing, mathematics, and communication), requiring an engineering process as well as the development of computational algorithms, which are encoded in the brains of honey bees. Such information-rich programs are not known to spring up through a series of small, purposeless evolutionary steps, with or without the benefit of something like natural selection. And there is nothing approaching a detailed proposal, credible or otherwise, for how these complexities might have developed in the case of honey bee communication and navigation.

Ant Navigation

IN A paper that appeared in the journal *Animal Behaviour*, Antoine Wystrach and Paul Graham from the University of Sussex wrote, "The remarkable navigational abilities of social insects are proof that small brains can produce exquisitely efficient, robust navigation in complex environments."[15] Their remark applies especially well to ants. Some ant species exhibit arguably the most impressive navigational abilities given their limited cognitive capacity and a brain size of approximately 250,000 neurons, one-fourth the size of the tiny honey bee brain.

The desert ants *Cataglyphis bicolor* and *Cataglyphis fortis* are particularly good examples. These hunting ants range relatively far away from their home nest, beyond the line of sight. While only a centimeter in length, they can travel more than a hundred meters. Without relying exclusively on visual landmarks or odor trails, they are still able to navigate

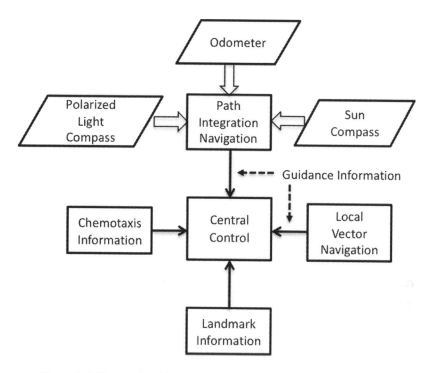

Figure 3.1. Desert Ant Navigation System

reliably and accurately. They can return to their nest in a direct route from any direction, even if they had taken a circuitous outbound route, as illustrated in Figure 2.3.

Desert ants employ three different methods of navigation: visual landmarks, vector memories of route segments, and path integration.[16] In addition, they use chemotaxis in close vicinity of a food source by the detection of odors. They also use a combination of sensor information sources for path integration navigation, including a sun compass, biological clock, and two forms of odometers.[17] The functional integration of this information in the navigation system is illustrated in Figure 3.1.

In path integration, the ant maintains a global vector, which points back to either its starting point or to a previously visited feeding site.[18] The path integrator uses information provided by two sources, a neural compass and a neural odometer. The compass information is derived from polarized light. As animal researchers Thomas Collett and Matthew Collett note, "The ants appear to keep track simultaneously of both the total distance that they have travelled from their start point (i.e. their global PI state) and the distance travelled along the current segment of their route."[19] The path integration function also requires the storage of compass and distance information in memory, for eventual computation of the route home.[20]

The odometer method of determining distance was confirmed in a clever experiment where the legs of ants were artificially both lengthened and shortened.[21] The experiment proved that ants somehow count the number of steps taken to determine distance. Experiments have determined that ants also employ another form of odometry based on optic flow (a method also used by bees and wasps). This method is based on keeping track of image motion during movement.[22] Collett and Collett note another indication of the functional integration of route segment odometry and global path integration. When both systems are active, "their outputs combine, and if the directions commanded by the two systems differ, the insect's direction of travel will be a compromise."[23]

Regardless of which navigation method is employed at any instant, *Cataglyphis* ants maintain the path integration information at all times. This means that even when the ant is using landmark navigation, if it loses the landmarks it immediately reverts to path integration, and continues on its intended route. These ants are also programmed to select the most accurate method, depending upon circumstances. When the environment is visually enriched, they will use landmark navigation; otherwise they will use path integration.[24] Additional research also has shown that desert ants can keep track of distance traveled and direction while moving backwards to drag food back to the nest.[25] (This ability to perceive their own movements or stimuli is called proprioception.) Such an adjustment is more challenging than it may sound, since as neurobiologists Sarah Pfeffer and Matthias Wittlinger at the University of Ulm (Germany) note, "Leg motion and proprioceptive input is not just a simple reversal of forward movement."[26] Accurate navigation while moving backwards is additionally impressive since the ants must deal with reversed sensory information, including the position of the sun and sunlight polarization pattern.

Another indication of their navigational ability is that if ants are moved and then released at a location away from the original route they "first search about the point of release, but wherever they come to hit the route, they rejoin it and follow it in the right direction."[27] Desert ants apparently calibrate their path integration routes by comparing the position with a known landmark, and adjust their route accordingly.[28] It has also been determined that despite the fact that path integration navigation accuracy theoretically degrades linearly with distance, as occurs in human-made inertial systems, path integration affords the ants surprisingly accurate information about the nest position even after wide-ranging foraging runs.[29] This is a further indication of the advanced engineering of their navigation system. And keep in mind, this is all managed in a brain that is only about one-fourth the size of a honey bee brain, and the skills are largely innate, not learned or taught. The hardware, if you will, comes installed with the software, right out of the box.

How the path integration function is encoded in an ant's brain remains a mystery. Different models have been hypothesized.[30] Some research points to the specific neural mechanisms that may be involved. There is also evidence that the same neural mechanisms might be shared among other insects.[31] While we do not know if any of the explanatory models are correct, taken together they strongly suggest that the mechanism must be complex and involve sophisticated algorithms. How did these complex programmed behaviors originate?

Monarch Butterfly Migration

EVERY YEAR an estimated 100–200 million monarch butterflies (*Danaus plexippus*) migrate two thousand to three thousand miles between the United States/Canada and Mexico.[32] While there are other populations of monarchs, including in western North America, South America, the Caribbean, and Australia, the population in eastern North America is the best known because of its amazing migration. During the migration, the butterflies lay their eggs on milkweed, where they then go through the larval and pupal stages. Milkweed is the only plant that provides food for the developing larvae. The butterflies are thus dependent on milkweed during their migratory route through the US.[33]

It typically takes up to three generations of butterflies to make the complete journey.[34] This means that the navigation information is genetically programmed. One of the unique aspects of the migration of the monarchs in eastern North America is that during their summer stay in Canada they occupy close to 400,000 square miles, while during their overwintering hibernation in Mexico they occupy less than half a square mile. As noted previously, they often migrate back to the same tree that their ancestor butterflies departed from in a mountainous region in Mexico. That means they must have an extremely accurate method of navigation to locate such a small target.

Monarchs navigate using a sun compass, and as previously described, this includes time compensation to account for the movement of the sun.[35] The circadian clock used in the process is embedded within the butterfly's antennae.[36] The sun's azimuth position is detected through the butterfly's compound eyes.[37] Researchers are only just beginning to decode the biological information required for these amazing feats. The genome of monarch butterflies has been decoded, including the genes related to the neurobiology and physical systems used for migration.[38] Comparisons of migratory monarch genomes with the genomes of nonmigratory monarchs has revealed that more than five hundred genes are involved in migratory behavior.[39]

A neuronal model has been proposed to explain how the time-compensated sun compass functions by integrating the azimuth-position information with the butterfly's internal circadian clock.[40] The theoretical model also explains how monarchs are able to maintain the southwest course in the fall, as well as the northeast course on the return migration in the spring. Further research is required to determine whether the model is correct. However, the model does help to underscore the programming complexity required of such a system and behavior. The model is also a good example of how engineering can be applied to analyzing complex programmed animal behavior.

At the same time, while this mechanism may explain how the system works, it does not explain where the information came from that defines the course flown by the monarchs, or the overall control and decision making. There exists no evolutionary model that satisfactorily explains its origin. That by itself does not prove that gradual evolution didn't produce such programming, but the lack of such a model should at least give the open-minded pause for reflection.

Monarchs are also able to continue migrating accurately when the sky is overcast and the sun compass is not available. This is possible since they use a magnetic compass as a backup source of navigation. They do this by sensing the inclination angle of the magnetic field to determine latitude.[41] The origin of the programming that allows for this also has not been explained in evolutionary terms.

Logically, we can approach the question as a decision tree of two possibilities. One, the satisfactory explanation for how such complex algorithms may have blindly evolved simply hasn't been found yet, despite considerable efforts to uncover such a process. Or two, no such process exists.

How to proceed? One is to take a never-give-up, never doubt approach. But that isn't how successful science is generally done. While there is a place for doggedness, science is ultimately about following the evidence, and the historical sciences, including origins science, is about seeking out the best explanation given the available evidence. If geologists proposed a series of purely natural, evolutionary explanations for the arrangement of rocks known as Stonehenge, and each succeeding explanation collapsed under scrutiny, over decades of investigation and conjecture one could insist that "absence of evidence is not evidence of absence." Or one could take the many failed attempts as a possible indication that Stonehenge was not formed by any blind, evolutionary process, and open oneself up to other possible causal explanations.

The point isn't that this must be the case with the migratory feats of monarch butterflies. The point is that it's illogical to never be open to the possibility that something other than blind evolution engineered a complex programmed animal behavior that remains stubbornly inexplicable in such evolutionary terms. Whether we have already learned enough about monarch butterflies to abandon blind evolution as an explanation for their navigational prowess is a question we will take up later.

Dung Beetle Navigation

Dung beetles are also amazing navigators. As their name implies, dung beetles feed on the excrement of other animals. They provide a valuable role in ecology by recycling nutrients and enhancing soil conditions. When a group of beetles detect a dung pile they each remove a small amount and roll it away from the pile (hence the name "ball-rolling" dung beetles). The purpose is to move the ball away as fast as possible from the other beetles and find a suitable location to bury it along with an egg; later the larva will feed on the ball of dung. In doing all this, the dung beetle also attempts to avoid other beetles, who might try to steal the dung ball. The way the beetle clears out as efficiently as possible is by moving away in as straight a line as possible.

Curiously, dung beetles do not use visual landmarks at all for this. Instead they employ a suite of other navigation sensors, including one that is especially striking: they use polarized light from the moon and stars.[42] Just as some animals can detect polarized sunlight and use it as a compass sensor, dung beetles can detect the polarized light from the stars in the Milky Way to use as the basis for a compass. Nocturnal beetles (*Scarabaeus satyrus*) use polarized starlight to navigate, but switch to the sun during the day.[43] Diurnal beetles (*Scarabaeus lamarcki*) also use the sun during the day but switch to using the moon at night.

Dung beetles have a special mechanism in their eyes that detects the light polarization vector.[44] Using starlight in this way is a particular challenge since it's a million times dimmer than sunlight. To manage it, the beetles have special neural mechanisms in the brain for detecting the lower light levels.[45] These mechanisms are in addition to the algorithms necessary for processing the information.

Turtle Navigation and Migration

LOGGERHEAD SEA turtles (*Caretta caretta*) start out life hatching from a nest buried in a sandy beach. The sea turtles we are familiar with in North America are born on the beaches of eastern Florida. Immediately upon hatching at night they instinctively crawl toward the ocean and start the swim out to sea. The cue they use to move toward the ocean comes from the light of the moon and stars reflecting off the ocean. That's the reason why local authorities ask people who reside in houses along the beach to turn off outside lights during this time, so the turtles won't become disoriented and crawl away from the ocean.

Loggerheads live most of their lives in the ocean. After they swim eastward into the ocean they eventually reach the Gulf Stream and then move into the Sargasso Sea.[46] The Sargasso is surrounded by the North Atlantic gyre, a circular current that takes the turtles on a circular migratory route around the North Atlantic, a route rich in food sources for loggerheads.

Females do not become sexually mature until they are twenty to thirty years old. At that time, they migrate back to the very same beach where they were born to lay their eggs. It has long been a mystery how the turtles are able to navigate back to the same beach. What makes this particularly remarkable is that they leave the beach right after hatching, and yet the beach location is somehow imprinted so firmly in their memory that they can find it twenty to thirty years later. They also continue to return to the same beach at two- to three-year intervals.

Research indicates that loggerhead turtles use the magnetic field to navigate.[47] Apparently loggerheads can detect both latitude and longitude from the magnetic field and thus may operate with a map sense.[48] Further, they swim in the proper direction, regardless of their location. In one experiment, "hatchling loggerhead sea turtles (*Caretta caretta*) from Florida, USA, when exposed to magnetic fields that exist at two

Figure 3.2. A Loggerhead Sea Turtle

locations with the same latitude but on opposite sides of the Atlantic Ocean, responded by swimming in different directions that would, in each case, help them advance along their circular migratory route."[49]

That laboratory experiment, conducted in a water-filled arena, simulated release of hatchling turtles on both sides of the North Atlantic (Puerto Rico and the Cape Verde Islands). The simulation replicated the magnetic information measured at the two locations, in effect inducing the turtles to "believe" that one of those places was their actual location. Their home destination is the east coast of Florida. In both cases the turtles swam in the proper migratory direction toward home. The hatchling turtles used in the experiment had never been in the ocean. So, they were not basing their navigation route on experience. This means that the navigation route is an innate programmed behavior.

There are three fundamental issues here to explain. One is the imprinting of the turtles' "home" beach location. Second is their ability to detect both latitude and longitude from the Earth's magnetic field and, with it, develop a map sense. Third is how the loggerhead sea turtles are programmed to swim in the most advantageous direction, depending upon their geographic location.

Trade-offs in Migration

ONE OF the interesting questions about migration is why some animals migrate and others do not, even among the same species. Many cases have been observed where only portions of a population migrate, and as Miriam Liedvogel, Susanne Akeeson, and Staffan Bensch of Lund University in Sweden note, crossbreeding experiments have shown migration to have a genetic component, with results that "suggest that many, if not all, birds have the genetic machinery to migrate."[50] Approximately 55–60 percent of bird species have a mix of individuals where some are migratory and others are not.[51] Blackcap warblers and robins are two examples.[52] Blackcaps in particular have a large variation in migratory patterns in Europe. Most, if not all, resident populations consist of a small fraction of migratory individuals, leading German ornithologist Peter Berthold to conclude that "this mix may facilitate the rapid evolution of adaptive migration patterns."[53]

What is at first somewhat puzzling is that there are significant advantages to not migrating related to survival and breeding success. Evolutionary biologists Derek Roff and Daphne Fairbairn observe that the risks of migration include being more susceptible to predation, and failing to find a suitable habitat.[54] In the case of robins, one study found that the survival rate for resident males was 50 percent compared to only 17 percent for migrants.[55] There was a similarly large advantage in mating success for residents.

So then, why migrate at all? Keep in mind that we don't have in view here animals that would freeze or starve if they didn't migrate. That is true of some migratory animals but not all. Rather, here we consider animals where there is a trade-off in cost versus benefit in migrating versus remaining resident. A simple trade-off compares the risk of surviving a migratory journey versus staying and starving due to lack of food during winter. Some have postulated that it is the dominant birds (usually adult males) that remain resident, while subordinate birds (usually females and juveniles) are forced to migrate.[56] However, that does not explain why those that migrate do not simply move to a different territory rather than migrate long distances.

Just as there is an advantage to maintaining diversity in other phenotype characteristics, diversity in migration is also beneficial. Research has shown that the proportion of migrants "varies from year to year and is determined by actual environmental conditions, such as breeding density or food availability."[57] One would expect that if all individuals in a population were subjected to the same environmental conditions and selection pressures, then Darwinian evolution would tend to eventually result in a population becoming either entirely migrant or non-migrant. However, that is typically not the case.

A potential reason to maintain diversity in migration habits is to enable adaptation to climate change and resulting changed habitats. We have witnessed some degree of global climate change over the past century, but it has been relatively modest compared to changes that have occurred over longer periods in the past. One needs only to go back to the most recent ice age, where much of North America and Europe was covered by glaciers. There obviously were great disruptions of animal habitats and migration patterns at both the onset and conclusion of this ice age.

Support for this theory comes from a study of blackcap warblers which suggests that current migration patterns developed very recently, probably following the glacial period. In some blackcap populations, "migration may have been lost again, after colonization of areas with mild winters."[58]

British ornithologist William Sutherland documented a number of changes in bird migratory patterns.[59] Experiments also have demonstrated that populations of migratory birds can become primarily resident and vice versa within a few generations.[60] Here we have in play populations that maintain diversity and plasticity in their migratory behavior, with natural selection impacting the survival of the fittest in a given climate context rather than being responsible for the arrival of a fundamentally new migratory behavior.

Evolution of Navigation and Migration Behaviors

So FAR we have considered in piecemeal fashion challenges facing evolutionary accounts of the origin of various navigation and migration behaviors. Now let's step back and examine the question more methodically. Figure 3.3 shows the minimal elements involved in navigation systems and migration behaviors and their specified characteristics.

So a lot goes into making a successful migratory animal. As Spanish animal migration researcher Francisco Pulido emphasizes, the migratory behavior in such creatures is a complex trait that includes a "high level of integration among single traits."[61] Population genetics models that underpin neo-Darwinian theory assume that all the information that defines these characteristics resides in the genome. A large number of genes are likely involved, as has been found in monarch butterflies. Pulido describes this as a genetic program where, in the case of migratory birds, the birds "possess an innate program that 'tells' them how fast to develop, when to leave the breeding area, how fast and in which direction to fly, and when to stop migrating."[62] Pulido insists that the origin of the migratory trait is the same as other complex traits, such that "whenever selection persistently favors the simultaneous optimization of multiple traits, genetic correlations among these traits will evolve."[63] However, for these behaviors to have evolved through neo-Darwinian mechanisms, they must have occurred through a gradual and incremental process, which is difficult to square with the "integration" nature of these traits noted by Pulido.

Problems facing the coordinated evolution of complex traits, such as those just discussed, are described by noted evolutionary researchers Günter Wagner and Vincent Lynch. "How novel traits arise in organisms has long been a major problem in biology," they write. "Indeed, the sharpest critiques of Darwin's theory of evolution by natural selection often centered on explaining how novel body parts arose."[64] A significant

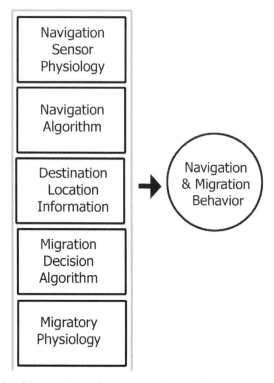

Figure 3.3. Navigation and Migration Control Elements

part of the problem is that "the overwhelming success of the population genetic approach during the decades following the modern synthesis all but sidelined the issue of innovation. It was more rewarding to calculate the variation of the existing rather than to puzzle over the origination of the unprecedented."[65] (This explanatory deficiency is one impetus for the investigation of epigenetic mechanisms as possible sources of novel traits. More on this later.)

Supporters of neo-Darwinian evolution insist that evolution is gradual. From Darwin down to the present, evolutionists in this tradition have acknowledged the prohibitive unlikelihood of evolutionary jumps involving multiple coordinated genetic mutations in a single go. That's because the improbability increases geometrically with each additional simultaneously coordinated mutation. Rolling a die and getting a six is a one chance in six. Doing it twice in a row isn't twice as unlikely; it's six times as unlikely, one chance in 6x6—that is, one chance in thirty-six. Doing it three times in a row is one chance in 216. The odds go down exponentially. A nightly poker player might draw a straight flush in a fair game of poker once in a blue moon (one chance in 72,192), but not three or four in a row. It's the same principle when considering multiple coor-dinated genetic mutations to create a new biological function in one fell swoop, except the odds are even more daunting than in the straight flush example. Research has found that the average rate for a single mutation in eukaryotes (multicellular organisms) is approximately one in a billion per generation.[66] Therefore, the likelihood of two coordinated mutations is one in one followed by eighteen zeros. This is why Darwinists properly insist that for evolution to work in producing novel form and function, the novelty must accrete one small mutation at a time.

A factor that limits the possible variations and subsequent selection is correlation between traits. When traits are correlated it restricts the number of variation combinations that are viable. Evolutionary geneticist Mark Kirkpatrick of the University of Texas (Austin) writes:

> Constraints become increasingly likely as the number of traits increase. When the number of traits is moderate to large, there are many more evolutionary forbidden changes than there are evolutionary permitted ones…. What do these constraints mean for adaptation? They frustrate selection from optimizing all traits simultaneously. Evolution cannot respond to immediate selection pressures that favor particular suites of traits. The evolution of one group of traits to their fitness peaks will drag other correlated traits away from their optima.[67]

There are a number of critics of adaptationism, even many who are otherwise supporters of Darwinism. The geneticist Richard Lewontin together with Stephen Jay Gould, both esteemed Harvard scientists, as-serted that the prevailing concept of adaptation is incorrect: "Too often, the adaptationist programme gave us an evolutionary biology of parts and genes, but not of organisms. It assumed that all transitions could occur step by step and underrated the importance of integrated developmental blocks and pervasive constraints of history and architecture."[68]

The next question is how the physiological and behavioral changes transpire that result in an adaptation. Variation and natural selection are extremely unlikely to produce complex programmed behaviors such as the characteristics associated with navigation and migration indicated in Figure 3.3. Assuming the information defining each of these elements resides in the genome, the development of the behaviors requires multiple coordinated genetic changes. This is because under Darwinian theory every small step must give an advantage; when multiple changes are needed before some advantage is conferred upon an organism, Darwinian evolution becomes impotent. Even Jerry Coyne, a leading "new atheist" and outspoken evolutionary biologist at the University of Chicago, admits this point: "Natural selection cannot build any feature in which intermediate steps do not confer a net benefit on the organism."[69]

Jerry Fodor and Massimo Piatelli-Palmarini make a similar argument in their assessment of the limitations of natural selection in gradual Darwinian evolution, writing, "Contrary to traditional opinion, it needs to be emphasized that natural selection among traits generated at random cannot by itself be the basic principle of evolution. We think of natural selection as tuning the piano, not as composing the melodies."[70] Their analogy indicates natural selection cannot be responsible for the origin of truly novel traits.

Cambridge University philosophy of science professor Tim Lewens discusses the interaction of different genes and how selection pressure affects their evolution, noting that in some cases, there can be conflicting trade-offs between genes:

> When an organism is troubled by these kinds of conflicting trade-offs, the complex interactions of selection pressures can stop any one of them from bringing any kind of complex adaptation into existence. This is a way of stating the familiar requirement that for complex adaptations to evolve by selection there must be a smooth series of variants, each of which is better than the last—not merely in terms of how it

performs its local function, but in terms of how it interacts with the whole organism.… Unless these assumptions are met, then selection, by changing frequencies of traits, can make the emergence of complex adaptations not more likely, but less likely.[71]

The major issue here is how all of these traits could have evolved in a gradualistic fashion to produce complex navigation systems and integrated migratory behaviors. In the case of birds, some evidence indicates that certain traits are correlated and interdependent. According to Francisco Pulido, it is currently not clear which aspects of the migratory trait are integrated and which could evolve independently.[72] Pulido's acknowledgment that some of the traits may not have been able to evolve independently is refreshing, since when one is committed to Darwinian gradualism, this can be a difficult possibility to acknowledge. But he appears to have gotten there by ignoring the difficulties this poses for Darwinian evolution. His assertion that natural selection will favor the optimization of separate complex traits in a gradual manner is simply an unproven assumption. In this case, there is no evidence that animal navigation systems evolve in a correlated manner with migratory behavior. That would involve significant and more-or-less simultaneous changes to two independent physiologies and two independent algorithms (four of the five elements in Figure 3.3).

All of these considerations mitigate against the gradualistic neo-Darwinian mechanism being an adequate explanation for these complex programmed behaviors.

Map Software Preloaded

ONE OF the biggest questions is what is the source of the programmed information that defines the destination or route during migration. Juveniles of many animals are born with a migration route already predefined. The previously described examples of the migration of bar-tailed godwits and bristle-thighed curlews fall into this category.[73] These migration routes must be somehow pre-programmed in the creatures since the routes are inherited rather than learned. That means it is not just the brain algorithm that controls the navigation, but also the actual route information, much as a map program on your phone requires both an algorithm and lots of specific information about the given route. This information must somehow be transferred into the animal's implicit memory.[74]

There is some evidence for a role of epigenetics in migratory behavior. As discussed previously, animals select when and where to migrate depending upon environmental conditions. That suggests an environmental effect on epigenetic mechanisms. An example is the migration of monarch butterflies in North America. Christine Merlin and Miriam Liedvogel write that the "switch in migratory physiology and behaviour and the reversal in flight orientation that occur respectively in the autumn and spring in response to environmental changes suggest an epigenetic control."[75] As for the genetic side of the picture, they comment, "The genetic architecture and molecular mechanisms that underlie the migratory phenotype, including flight direction/orientation and timing of their migration, remain poorly understood."[76] If both genetic and epigenetic mechanisms are necessary to control a behavior, this suggests that multiple coordinated changes are necessary for a trait before it can confer some advantage—precisely the sort of multi-component trait that challenges a Darwinian explanation. One might wish to appeal to some law-like cause giving rise to these capacities, but the diversity of migratory behaviors within the same animal populations mitigates against any such explanation.

Animal Navigation in Light of Aircraft Navigation

As we have seen, a wide variety of animal navigation and migration patterns strongly suggest complex programmed behaviors. The migration and navigation strategies used by most animals are far more sophisticated than initially supposed when scientists first started studying them. The more scientists learn about them, the more complex they appear in many animals.

One way of demonstrating the level of engineering involved in these behaviors is by examining modern aircraft navigation systems. As shown in Figure 3.4, aircraft navigation systems use multiple redundant sensors (inertial, GPS, ground-based systems) along with other essential components.

The Figure 3.4 illustration is, of course, a gross oversimplification. Having been personally involved in the engineering design of several systems, including aircraft navigation systems, I can attest that a structured engineering process is essential. The process must be top-down, where the overall concept must first be defined, as illustrated in Figure 3.5.

The process is as follows. The first step is to define the overall goal, including the purpose of the system. The second step is to develop a concept for implementing the system including the primary functions. Next is performing an analysis of potential design options. This includes assessing the options and evaluating the trade-offs, which for man-made systems include performance, complexity, and cost. Once a design option is chosen the next step is to define the specific requirements. Following that, the system can be manufactured.

There are numerous reasons why this process must be top-down and structured. One is that since it is a complex system there are a lot of interdependencies, and therefore the design requirements must take this into account to ensure the components and operation function coherently. If all of the functions are not integrated correctly, system performance is significantly degraded.

The overall design of aircraft navigation systems requires thousands of hours of design work and development by engineers. The goal is a system that provides the optimum navigation information to the pilots by selecting the navigation source that provides the best performance for specific phases of flight—takeoff, en route, oceanic, airport terminal area, approach, and landing. The functions include the various sensors that provide this information, each of which is itself extremely complex.

As with all modern systems, it includes a combination of hardware (radio receivers, computer processors, cockpit displays) and software containing the computer algorithms for processing aircraft position data, navigation selection logic, map and route information, and display

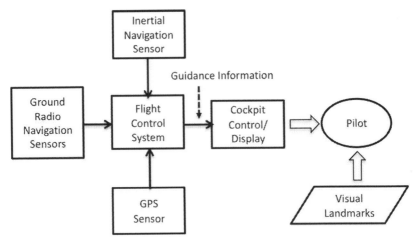

Figure 3.4. Aircraft Navigation System

interface. All navigation methods require an integrated and coherent combination of physical elements, programmed algorithms, and other related information. The same is true for all migrating animals.

Referring to Figure 2.2 regarding animal navigation systems, recall how all of these elements interact. Or to take one of the specific examples from above, we see this in desert ants (Figure 3.1). While the sensors, brain, algorithm, and physiology can be viewed as separate subsystems, they must work as an integrated system. Recall that desert ants employ three methods of navigation. Notice the similarity to aircraft navigation systems in Figure 3.4. The odometer and polarized light compass in desert ants are the information sources for path integration, while landmarks are the other information source. The central control function uses this information, in addition to the cues available from external conditions, to make a programmed decision of which source to use and to compute the correct navigation route. As we can see from all of this, the control function is a complex algorithm.

And it only gets more complicated. Referring again to Figure 3.3, the information defining these elements likely resides in different parts of the genome. Further, the information defining these elements is of a very different nature. For example, genes that control physiology bear

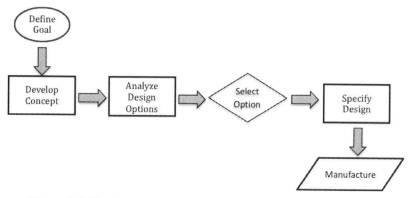

Figure 3.5. Top-Down Engineering Process

no relationship to genes that define navigation or migration algorithms. Even the genes that determine sensor physiology (compass sensors, etc.) are very different from genes that determine migratory physiology (flight characteristics, etc.). All told, the development of navigation and migration behaviors requires the independent origin of the physical traits and information necessary for five separate groups of genes and other genetic information in the genome. As discussed above, the likelihood of obtaining even a few coordinated genetic changes is very low. The likelihood of obtaining an unknown (but likely large) number of coordinated genetic changes in five different parts of the genome is extremely improbable. Also required is the development of novel traits (of which there are numerous related to navigation and migration), which will be discussed in Chapter 4.

I wrote an initial draft of this section shortly after there were two incidents in the news where commercial aircraft landed at airports that were not the intended destination. This is an example of pilots either failing to refer to their alternate navigation systems, or actual failure of the system. In either case, the incidents illustrate how crucial it is to have backup navigation strategies, and how challenging it is to program systems that consistently work properly. It is a fundamental concept in engineering that when the availability and reliability of a system is critical, then it's best to design a backup system. A given cue may not be available at certain times, and at other times it may give ambiguous information. Having an alternate source of information can also help detect erroneous cues. We find this rule of thumb followed in desert ants and in many other migratory creatures.

Those who program computer algorithms for a living are especially well situated to appreciate just how complex an algorithm would need to be to function as well as the desert ant's navigational algorithm. In this case, the algorithm consists of several logical decisions based on the information detected. Once the algorithm makes a decision concerning which source of navigation to use based on incoming data and environmental conditions, the control function then must compute the course. In some instances the course is computed based on combining the information from two navigation methods. When the ant is using the path integrator it keeps track of its movement through the odometer and the angular movements determined by the compass, which are both stored in memory. It then can compute the direct path to its home nest. This is a complex process that involves trigonometry.

While we can write a reasonably simple segment of computer code to do this computation, in this case it must be programmed into the ant's brain and integrated with the many other essential elements of the navigation algorithm. The question is twofold: (1) How can a trigonometric mathematical computation be programmed into the brain of an ant through a neo-Darwinian process of genetic mutation and natural selection? The programming, keep in mind, likely involves a neural circuit, and one of considerable sophistication. (2) How could a neo-Darwinian process manage to do so while simultaneously building other essential subsystems of the larger integrated navigation system? The trigonometric calculation alone would seem to be difficult to evolve in a stepwise manner, but without these other subsystems in place, the most beautiful algorithm in the world for calculating a trigonometric mathematical computation is useless to the ant, and therefore not available for being seized upon and preserved by natural selection. This raises the chicken-

or-egg problem: Which evolved first—the physical systems or the behavior algorithms?

Serial gradualistic evolution does not seem plausible, as each characteristic by itself is not useful. On the other hand, simultaneous evolution of all of the physical characteristics and behaviors is not plausible due to its extreme improbability.

Or consider another programming challenge. There is evidence that birds can use one type of navigation cue to calibrate another one. For example, an experiment with *Catharus* thrush songbirds indicated that the magnetic heading they use during migration is calibrated daily relative to the solar azimuth at sunset or twilight, rather than maintaining a fixed orientation to magnetic north.[77] For animals that migrate relatively long distances, multiple sources of navigation information make sense from an engineering perspective. This too has a parallel with aircraft navigation systems. In some integrated systems, one sensor is used to calibrate another.

Consider also the complex algorithms needed to compute navigation routes. These include the algorithms needed for path integration, computation of great circle routes in long-distance migration, and computation of time compensation using a sun compass. Neo-Darwinian evolution would be hard pressed to explain the origin of these algorithms one at a time, much less in concert and together implemented in brain neural networks.

It is far from clear to anyone how even one of the navigation strategies could have evolved through a neo-Darwinian process, let alone multiple integrated strategies. This difficulty, moreover, doesn't arise from a failure to think hard about the problem. Quite the opposite. The more one understands about engineering and the challenges in engineering complex functional systems—in this case, navigation systems—the more implausible a neo-Darwinian origin scenario appears. In other words, this is an inference based not on ignorance but on engineering know-how.

4. Complex Programmed Societies

Instinct knows everything, in the undeviating paths marked out for it; it knows nothing outside those paths.[1]

— J. Henri Fabre

SOCIAL INSECTS MAKE FOR A GOOD CASE STUDY OF COMPLEX PROgrammed behaviors (CPBs). Consider the ant. All species of ants (family Formicidae) live in groups, some small and others quite large. As with many social insects, ants exhibit several complex behaviors associated with social group lifestyles, including agriculture, territorial wars, slavery, division of labor, castes, consensus building, cities, and a symbolic language.[2] And all of these complex social behaviors appear to be programmed, since the ants master the behaviors almost immediately, before there is time to have learned them.

In general, there are many benefits to group living, including defense against predators, foraging/hunting, and care for young.[3] Many types of animals exhibit social behavior, from invertebrate insects to the most advanced vertebrates, such as primates. For convenience these have been divided into three categories, moving from the less socially complex to the more socially complex: incipient sociality, basic eusociality, and complex eusociality.[4] *Eusocial* describes species where usually a single female or a caste of females is reproductively active while others care for the young, gather food, and provide protection. Biologist and insect social behavior researcher Karen Kapheim comments that eusociality is "one of the most complex social behaviors known to animals."[5] Edward O. Wilson, who pioneered the field of sociobiology, a discipline that attempts to answer questions about the ultimate causes of social behaviors,[6] describes the most advanced instances of eusocial organization as a "superorganism." He and his colleague, Bert Hölldobler, define this as "a colony with many of the attributes of an organism but one step up from organisms in the hierarchy of biological organization."[7]

Given the sophistication of these eusocial colonies, it's curious that they are almost entirely restricted to insects with their tiny brains. (Naked mole rats are the only known vertebrate animals that exhibit eusocial behavior.)

The most familiar types of insect social groups are grouped within order Hymenoptera, which includes ants, bees, and wasps. While all species of ants are social, most bees are not. Even so, there are about one thousand species of bees that are social, the one most familiar being honey bees.[8] Also, some bee species are "facultative," meaning they can express different forms of social organization.[9] Among eusocial wasps (mostly belonging to family Vespidae) there are social groups of paper wasps, yellow jackets, and hornets. Like ants, all species of termites (infraorder Isoptera) are social.

Of Colonies and Castes

MOST INSECT societies are composed of several castes, where groups of individuals perform different functions. The primary division is between individuals that reproduce and those that do not. Typically, the queen is the only female that reproduces. Those that do not reproduce are primarily workers that perform most of the tasks of the colony. This includes building a nest, foraging for food, feeding and taking care of the brood, and defending the nest against predators and other attackers.

In the case of honey bees, besides the queen, the other two castes are the workers and drones. The drones are males who mate with the queen, but perform little other work in the colony. An interesting side note is that the male drones die shortly after mating with the queen. Workers are females that cannot reproduce, so they have zero "direct fitness" in Darwinian terms.[10]

The differences among the castes are usually behavioral, but in many instances, they are also physical. In paper wasps, the queen and workers are all physically similar. On the other hand, in vespine wasps the queen is larger than the workers. In most species, the queen and female worker castes are genetically identical. Environmental differences determine whether a female becomes a worker or a queen. Several variables have been identified as determining when females become queens. These include larval nutrition, temperature, egg size, age of current queen, and self-inhibition of the caste.[11] In ants, honey bees, and termites the presence of a queen inhibits the production of a new queen. Differences in caste, according to insect behavior researchers Amy Toth and Sandra Rehan, are a "classic example of phenotypic plasticity, and environmental caste determination (relying on gene expression rather than heritable, genotype differences) is considered the rule in most eusocial insects."[12] This is also an example of epigenetic control of phenotype and behavior.

New colonies are formed when prospective queens and males leave an existing colony, mate, and settle into a new location. The dispersal flights in some species can be spectacular, where thousands of males and queens meet. There are several variations in this basic life cycle theme, including having multiple queens, or even a king in the case of termite colonies.

Wasps of genus *Polistes* exhibit relatively simple social behaviors. Their colonies are small, usually less than one hundred individuals, and have only one or a small number of females.[13] In most cases they have only one queen, who differs little physically from the subordinate workers. The queen limits the reproduction of subordinate females through physical aggression. There is not a clear division of labor as in the more complex colonies. On the other end of the spectrum are driver ants and leafcutter ants. As biologists Brian Johnson and Timothy Linksvayer explain, these ants have complex social structures marked by "inflexible castes governed by elaborate systems of communication (social physiology) that facilitate collective decision making in every domain of colony life."[14]

Other complex programmed behaviors essential to the functioning of eusocial insect colonies involve architecture, communications, navigation, mating, and foraging.

There are four common features of animals within social systems: 1) They are highly sensitive and responsive to environmental and social information. 2) This information is transmitted within individual animals through one or more sensory pathways. 3) The signals are processed and integrated in particular circuits of the brain. 4) The animal's internal state controls its ultimate behavior.[15] Figure 4.1 illustrates how these elements interact to produce social behavior.

The diagram illustrates how many different elements are involved in social behavior tasks and the complexity of their integration. The critical elements of these features are the reception of information by sensors and processing of social signals, and how the resulting behavior of individuals in a colony changes.

Who calls the shots in these insect colonies? There is no central control of decision making. Instead, there is a distributed network among all members. In the case of honey bees, the workers perform various tasks, including shaping the comb, tending to the brood, cleaning cells, foraging, making orientation flights, patrolling, and eating pollen.[16] Determining when these tasks are performed is based on both communication

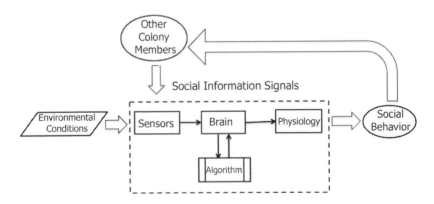

Figure 4.1. Social Behavior System

of social information signals within the colony and environmental cues. The environmental cues include nest temperature, shape of the cells, moistness of larvae, and degree of crowding at a food source.[17]

The individual behavior is largely innate rather than learned and thus appears to be based on algorithms programmed in the brain. The algorithms process the information from the environment and from colony members to make a decision as to the appropriate behavior. The behavior of each individual in turn can influence the behavior of other members of the colony—algorithms affecting algorithms in a recursive process.

Division of Labor

ONCE ONE becomes aware that much of social insect behavior is programmed, it may be tempting to assume that such programming means inflexible and repetitive behavior. Not so. The behavior of each individual insect often involves quite flexible decision-making based on multiple variables.[18]

In many cases individuals are programmed to perform multiple tasks, guided by a flexible task-allocation system that affords the colony the ability to rapidly respond to shifts in demand for specific tasks.[19] The number of workers engaged in the various tasks of the colony changes continuously, partly depending on the number performing other tasks. This indicates that the process of real-time task-allocation is more complex than originally thought.

Individual colony members do vary in their predisposition to perform certain tasks. In many social insects the predispositions change with age (called polyethism), while other predispositions are based on size (alloethism).[20] Usually it isn't difficult to uncover the adaptive benefit of such arrangements. Consider, for instance, foraging insects. They incur a much higher risk of mortality than do workers inside the nest, with one study of a colony of the hornet species *Vespa orientalis* finding that within the first fourteen days of adult life, inside workers had a mortality of 8.8 percent, while foragers had a mortality of 42.5 percent.[21]

Foraging insects are typically older. This division of labor serves to minimize the total number of life days lost "in the field," thereby benefiting the colony.

Various models have been proposed to explain the mechanisms of task selection for division of labor.[22] There is no consensus as to exactly how task selection is mediated in social colonies, though some things are clear. The division of labor, for instance, is often enforced through dominance hierarchies. In some cases, the hierarchy is based on age, as described above. One study found that the likelihood of ant foraging was related to corpulence (weight), where the lighter members tended to engage in foraging more than heavier members.[23] In other cases, the roles are determined by competition for reproductive rights, with the more dominant individual having the higher rank.[24] These pecking orders quickly get sorted out and, according to Joan Strassmann and David Queller, "the individuals usually do seem to act toward a common goal."[25] The result is an extremely efficient organization that would be the envy of many a corporate manager.

Social Information Communications

An important part of social behavior systems is the feedback among colony members that influences the behavior of other members. In the case of insect colonies this is largely done through pheromones.

The generic term for a substance that functions as a communications signal between organisms is a semiochemical. Most insect social behavior relies to some degree on them.[26] Hölldobler and Wilson identify twelve functional categories of communications of social insects, most of which are chemical: alarm, attraction, recruitment, grooming, feeding, exchange of fluids and solid particles, group effect, recognition of nestmates, caste determination, control of other individuals competing for reproduction, territoriality, and sexual communication.[27]

The importance of semiochemicals in insects is indicated by the multiple sensors used for detecting them. They include antennae, maxillary palps (mouth parts), and gustatory (taste) cells on the proboscis, legs, wings, and genitalia.[28] Most pheromones are processed by the receiving animal's olfactory system, with a small percentage processed through the taste system.[29] In many cases pheromones are derived from compounds that have other functions.[30]

Ants provide a good illustration of the multiple uses of pheromones in foraging, kinship discrimination, mating, and territoriality. In some species of ants the same gland is used to produce pheromones to (1) recruit foragers, (2) mark trails, (3) signal for help in carrying large prey, (4) orient toward home, and (5) mark a home range.[31] Hölldobler and Wilson comment that ants are the "insect geniuses of chemical communication."[32]

Ants employ numerous compounds in pheromones, with single compound pheromones being a rarity. The carpenter ant (*Camponotus ligniperda*) secretes at least forty-one compounds, and the weaver ant (*Oecophylla longinoda*) over thirty.[33] Some of the pheromones are generic, in that ants of any species will respond to them, while others are species-specific. Homing signals are often specific to individual colonies. A combination of pheromones is also used to identify specific colonies in some species. These create a unique "colony odor." Hölldobler comments, "To achieve such a high degree of specificity, such recognition labels must be rather complex, multicomponent signals."[34]

Some of the complexity in chemical communication involves the decision-making of the signal receiver. Evolutionary biologist Tristram Wyatt explains: "The neural circuits can be thought of as acting like digital logic 'AND' gates: if a component is missing or at the wrong ratio, the stimulus does not go higher in the brain. Conversely, the circuit gives a 'STOP' if there is activation of olfactory sensory neurons sensitive to a pheromone component of the wrong species."[35] That describes the simplest application of chemical signals. The full range of signals and associated behaviors is much more complex, indicating that the processing of this information involves complex algorithms.

Altruism

A KEY element of social behavior is the apparent altruistic behavior of individual members of social groups. An individual acting in an altruistic manner will act in a way that helps the survival of other individuals, while diminishing the chances of its own survival or that of its potential offspring.[36] We see this among social group animals that cannot reproduce and thus cannot pass their genes on to offspring. Examples are the sterile worker and soldier castes of some social groups. A dramatic example of altruism is the behavior of termites who in defending their nest will sacrifice themselves in fighting off an attack of army ants even when the soldier termites are clearly doomed to die if they don't retreat. Such altruistic behavior would seem to run contrary to the Darwinian paradigm and the "selfish gene" view espoused by Richard Dawkins and others, which suggests that natural selection is constantly pushing animal life toward individual forms that seek to maximize the likelihood of their own survival or the number of their own offspring. (More on this subject later.)

Advantages of Social Systems

THE EFFICIENCY of social insect colonies has made them extremely successful. Due to their superior performance in expanding their population, social insects greatly outnumber solitary insects such as cockroaches, grasshoppers, and beetles.[37] This is despite the fact that social insects represent only about 2 percent of the approximate one million species of insects.[38] Superorganism colonies range in size from hundreds of thousands to millions of individuals. An example was documented in a study of a region in the Amazonian rain forest where it was found that social insects (wasps, bees, ants, and termites) make up about 80 percent of insect biomass there and more than 20 percent of the animal biomass.[39] Wilson observes that animals in land environments are dominated by species with the most complex social systems.[40] A further indication of how successful social insects have been is that there isn't any evidence that major groups of social insects have gone extinct.[41]

Bees

EUSOCIALITY HAS developed independently at least six times in bees, more than in any other group of social animals.[42] The degree of social organization among these creatures is staggering. Bee expert Jürgen Tautz goes so far as to compare a bee colony to a single organism. "A honey bee colony is equivalent not only to a vertebrate, but in fact to a mammal, because it possesses many of the characteristics of mammals," he writes.[43] Such characteristics include protection of offspring in a precisely controlled environment, maintenance of constant temperature, learning ability, and low rate of reproduction. The latter is related to the fact that there are only a small number of reproductive females, which are produced only when they are needed. This happens when the worker bees feed a special diet to larvae housed in queen cells in the comb.[44]

The division of labor in bees is related either to age or size. Bumblebees use size as the basis for labor divisions. According to entomologists Guy Bloch and Christina Grozinger, the largest members of such colonies are nine times bigger than the smallest, and "large individuals are more likely to perform foraging activities, and may start foraging as early as at their first day as adults, while small bees tend to perform in-nest activities."[45] Recent studies show that large workers are better suited for foraging activities because they have more sensitive sensory detection systems.

With honey bees (as well as with ants) there is abundant evidence of innate developmental programs for physiology and behavior related to age and in the service of an age-related division of labor.[46] Also notable is the fact that the bees can perform various tasks in the division of labor, including foraging (which requires navigation and an ability to memorize numerous cues about the flowers), finding new comb locations, building the comb, and cell cleaning and repairing. We see adaptability in action, for example, when young honey bees transition to foraging if the older bees are removed from the colony.

Their ability to perform all of these tasks implies that one of two options must be true. One option is that the animal has the cognitive capacity to make decisions and carry out tasks that it determines are required. "The cognitive implications of a jack-of-all-trades approach are substantial," the Goulds comment.[47] The second option is that all of the tasks are pre-programmed, along with the decision-making apparatus to determine which task is required. The former option is not likely since bees have extremely small brains and limited cognitive capacity. That leaves the second option as the live alternative. However, that still leaves a problem in managing the behavior of thousands of individuals in a colony.

Individual bees do not have an overview of the needs of the colony in terms of supply and demand of food. However, from experiments it is known that a honey bee colony optimally adjusts the number of active foragers, which is achieved through some sort of decentralized mechanism.[48] Based on his extensive study of honey bees, Cornell University biologist Thomas Seeley points to "special systems of communication and feedback control."[49]

Honey bees also engage in several behaviors to control the environment in the hive, which is critical to raising the brood. Temperature is maintained between 93°F and 96°F, even as the outside air can range from -20° to 120°F.[50] This is accomplished by the bees in the colony, which either raise or lower their metabolism. Raising metabolism increases the temperature of the bee bodies, thus raising the temperature in the hive. The reverse is true when lowering metabolism. They also regulate the levels of carbon dioxide through increased ventilation when CO_2 levels get too high. They even fight off fungal infections by raising the internal temperature.[51]

On a side note, the life of honey bees isn't just good for the bees. Pollination by animals is crucial to almost all terrestrial ecosystems. And honey bees, along with bumble bees and some solitary bees, are crucial for pollination.[52] About 40,000 flowering plant species are highly dependent on honey bees.[53] Bees thus provide an essential function in maintaining the diversity of plants.

Ants

THE TERM "superorganism" was first applied by the famed Harvard entomologist William Morton Wheeler in 1911 based on his work with ants.[54] Hölldobler and Wilson define a superorganism as "a colony with many of the attributes of an organism but one step up from organisms in the hierarchy of biological organization."[55] All ant species are social and ant colonies exhibit several behaviors that only work as part of a group effort. The various ant species are extremely diverse in their diet and in how they organize socially and specialize in particular lifestyles. Specializations include farmers, architects, engineers, and soldiers. (In each case the specialist species also performs other essential behaviors.) Hölldobler and Wilson label leafcutter ant colonies as "Earth's ultimate superorganism."[56] Their diets range from seeds, leaves, and fungus to animal prey and excrement from aphid insects.

Division of labor among the worker castes is essentially universal in ants. The roles generally appear to change as the ants age, although they sometimes will switch back. Younger ants work inside the nests, tending to the brood, while older ants work outside the nests foraging for food and participating in defense.[57]

In a minority of ant species (and termites) there are sub-castes among the workers, which does not occur at all in bees and wasps. In some species of ants, the differences in the sub-castes are physically dramatic— differences that are used for highly specialized tasks and behaviors. An example is sub-castes that have large heads used in breaking open seeds. Several soldier ant sub-castes have weapons honed for combat. They include fierce-looking mandibles shaped like sabers and used to attack other ants or prey. Some soldier ants also have heads with horns or spikes. Turtle ants (*Cephalotes varians*) have a disc head structure used to block the nest entrance.[58]

Another group of ant species with a sub-caste of large-headed soldiers is the genus *Pheidole*. Interestingly, they have less defensive capabilities, but make excellent attack soldiers. They also do not possess ovaries, and thus cannot reproduce. Hölldobler and Wilson describe this sub-caste as a "throwaway" caste because they are "small, light, cheaply manufactured, and short-lived."[59]

The physical diversity among ants is matched by their remarkable behavioral diversity. Hölldobler and Wilson note the following behaviors in the species *Pheidole dentata* alone: they attend the mother queen, they groom eggs, they groom microlarvae, they "roll and carry eggs and microlarvae," they help adults emerge from pupae, they groom pupae, they roll and carry both pupae and mature larvae, they groom larvae, they "exchange oral liquid with mature larvae," they retrieve prey, they stand guard over the nest entrance, they guard the food stores, they forage, and they excavate and defend the nest.[60] Quite an impressive list.

As with bees, these behaviors, and the coordination of all these behaviors in a smoothly functioning ant colony, are particularly remarkable given the tiny size of ant brains. It is clear that these behaviors are programmed, but also involve decisions concerning the time, place, and external conditions in which to perform them. All of these behaviors and decision rules point up the need for a complex control algorithm.

Ant colonies can grow to mind-boggling size. One colony in the Hokkaido region of Japan was measured to cover 675 acres with over 45,000 interconnected nests.[61] It was estimated to have over 300 million workers and a million queens.

Most ants live for only a few weeks, though many queens can have a considerably longer lifespan. The queens of most species usually live at least five years, while some kept in captivity have lived for twenty to thirty years.[62] Queen ants are also incredibly fertile breeders. Leafcutter ant queens give birth to as many as 150 million workers, and two million to three million of a queen's offspring may be alive at one time in a given colony.[63]

As for the males, only a small number are fortunate enough to mate with a queen. They will also die within hours or days after leaving the home nest.[64]

Let's take a closer look at perhaps my favorite ant genus, the leaf-cutters (*Atta*). "If a congress of naturalists were to gather to choose the seven wonders of the animal world," write Hölldobler and Wilson, "they would be compelled to include the bizarre and mighty civilizations of the *attine* leafcutters."[65] *Atta* colonies live in symbiosis with and farm fungus for their sustenance. And by "farm," I mean farming in a very familiar sense. For example, the ants fertilize their fungus crop much as human farmers fertilize. Also, these ants use their razor-sharp mandibles to cut leaves from trees to provide the fungus a type of mulch. Equally impressive if more subtle is the skill and efficiency with which they carry out their cutting and mulching efforts. They maintain a particular position that maximizes cutting efficiency. And they cut the leaves to a size and weight optimal for getting the most bang for the effort—not so small as to waste trips, not so large as to waste time and energy on an unwieldy load.

Ants also have been observed working in tandem to cut twigs, much as two people collaborate using a crosscut saw to fell a tree. The larger worker ants do the cutting while the smaller ants transport the leaves back to the nests in an assembly line. That is a sight to see, as the ants follow trails back to the nest that are maintained to maximize transport efficiency, carrying the leaves in a fashion that gives them the amusing appearance of holding umbrellas (Figure 4.2).

The leafcutter ant trails can be as long as 150 meters,[66] with the transport sometimes accomplished through a form of "bucket brigade" where the ants carry the leaves for a distance and then pass them off to other ants. These brigades consist of up to five separate ant groups.[67] Some theorize that the bucket brigade functions to speed up communications among the foragers. The purpose would be that as the foraging conditions change (e.g., quality of the material) the ants can respond more quickly.[68]

Figure 4.2. Leafcutter Ants at Work

The ants also do various things to prepare the leaves, including stripping away the outer waxy layer and shredding them. The fungi are saprophytic, meaning they break down the organic material, including cellulose, thereby turning it into edible food for the ants. Ulrich Mueller, a biologist at the University of Texas at Austin, observes that the more he studies these fungal gardens of the leafcutter ant, "the more analogies I find between human agriculture and ant agriculture."[69]

Atta colonies usually have only one queen, along with hundreds of thousands up to several million workers. It has been estimated that leafcutter ant colonies can harvest more than a thousand pounds of plant material in a year—close to fifty thousand square feet of leaf area.[70]

Another unique aspect of their farming behavior is their ability to maintain optimum conditions for the fungus. They are choosy about the types of leaves they bring back to the fungus, discriminating based on the physical and chemical conditions of the leaves. This includes rejecting plant material that contains chemicals that are harmful to the fungus.[71]

The ants are also selective in maintaining the same variety of fungus. Leafcutter ants meticulously prevent different strains of fungus from mixing with the primary one for that colony. One study found that "*Acromyrmex* worker ants actively discriminate between fragments of their resident fungus and genetically different symbionts, which confirms predictions that the ants have a strong short-term interest in maintaining genetically pure gardens."[72] A possible functional reason for such behavior: the presence of an alien fungus may in many cases reduce productivity, harming both the resident fungus and the ants.[73]

There are various other ways the ants maintain the health of the fungus. These include producing growth hormones and secreting antibiotics to depress competing fungi and microorganisms.[74] Hölldobler and Wilson comment that the symbiotic relationship between leafcutter ants and fungus "should be viewed as a highly integrated superorganism that is more than the sum of its parts. The ants' division of labor and much of their social behavior are shaped by the details of this symbiotic relationship."[75]

Leafcutter ants have a significant positive impact on ecosystems, as they produce large amounts of organic matter in the soil. One study compared the nutrient level in ordinary leaf litter with the refuse produced by leafcutter ants. The results indicated that the ants' refuse contained forty-eight times the level of nutrients found in typical leaf litter,[76] including greater concentrations of two common ingredients in chemical fertilizers on sale at gardening stores, calcium and potassium. The increase in nutrients causes increased root production, which in turn leads to more tree growth.

Termites

THE MOST advanced social termites (infraorder *Isoptera*) have colonies that can reach into the millions. One factor that enables this is that the queen grows to an enormous size, reaching a length up to four inches, and having an egg-laying rate that exceeds that of all other animals.[77] It is only through the social behavior of these species that they can construct spectacular nests and mounds, which I will describe in the next chapter.

We think of termites as pests because of their taste for wood, obviously undesirable if the wood they're eating happens to be your house. However, they fulfill an important ecological role in many regions, predominantly in sub-Saharan Africa and Southeast Asia, representing up to 75 percent of insect biomass in these tropical areas.[78] Their ecological role is to eat dead plant material, converting it into material useful to other animals. Only a limited number of vertebrate animals can manage this, mainly ruminants such as cows. Other insects fulfilling this role include cockroaches, crickets, flies, and beetles. However, termites are the only social insects in this group, and their output is extraordinary. They can decompose up to 90 percent of dead plant material in areas where they are abundant. There are thousands of species of termites, and they are so diversified that "specialist genera exist for virtually every type of plant detritus, wood, humus, and dung, in every stage of decay, in virtually every type of temperate, subtropical, and tropical habitat."[79]

All termites are eusocial, and the most advanced have multiple caste systems. There is also a group (subfamily Macroterminitinae) that farm fungus in a fashion similar to leafcutter ants. This is an example of a convergent behavior between ants and termites,[80] though the symbiotic relationship with the fungi isn't the same in the two cases. In Macroterminitinae, older termite workers collect plant material and transport it to the nest. Younger workers eat the plant material along with *Termitomyces* fungal spores. This spore-plant mix is only partially digested, with the help of gut bacteria, and then defecated, providing a new layer of fungus. The *Termitomyces* then grows rapidly on a diet of the spore-plant mix, after which older termites consume the new-grown fungus. At the end of this process nearly all of the organic matter has been broken down.

A recent analysis of the genomes of one Macroterminitinae termite species and of the symbiotic bacteria suggests that the converting of biomass in termite mounds is primarily accomplished through cooperation between the termite's domesticated fungus and a specialized bacterial community.[81] The University of Copenhagen's Michael Poulsen and his co-authors characterize it as a unique form of relationship based on a "simultaneous tripartite life-history transition: insects becoming farmers, fungi becoming crops, and gut microbiotas adopting largely unknown complementary roles."[82]

In Darwinian terms this relationship is assumed to have developed through coevolution. However, this requires the coevolution of three entirely separate genomes (termite, fungus, and bacteria) to foster the symbiosis. This is an extremely complex relationship that involves numerous genes in each species. The coordinated Darwinian evolution of all the required genes in three independent genomes is a highly improbable scenario.

Social Behavior Plasticity

PLASTICITY (THE capacity of organisms to change or modify) is a characteristic that has been proposed as a potential explanation for the origin of novel traits. For instance, theoretical biologist and US National Academy of Sciences member Mary Jane West-Eberhard suggests phenotypic plasticity can significantly influence evolution in several ways.[83] (A phenotype is the observable characteristics of an organism. Its genotype is its particular set of genes.) Obvious examples of phenotypic plasticity include the variations in the size of animals, which depends upon not only genetics but also environmentally influenced factors such as diet and exercise.

The concept extends to behavioral habits. Darwin suggested that new species well might arise based on individuals developing different habits or behaviors, even while the morphology (the form) of such a line changed very little, as in the case of some varieties of woodpecker.[84] While most woodpecker species forage for insects in the trunks of trees, there are some that eat fruit, and others that reside in areas where there are no trees.

Emilie Snell-Rood, an ecologist at the University of Minnesota, describes two forms of plasticity: developmental and activational. The developmental form occurs "where a genotype expresses different behavioural phenotypes in different environments because of different developmental trajectories triggered by those environments."[85] Developmental plasticity includes learning as well as physiological changes, such as change to neural networks and sensory systems. Snell-Rood defines activational plasticity as "differential activation of an underlying network in different environments such that an individual expresses various phenotypes throughout their lifetime."[86]

Social insects exhibit both developmental and activational plasticity. This includes their response to environmental factors in the development of different castes, and their response to environmental conditions and other social factors in the division of labor.

The benefit of plasticity is that most environments tend to vary a lot, so no one behavioral phenotype will prove consistently optimal; behavioral plasticity allows an organism to quickly adapt to track changing environmental conditions. Such plasticity is no simple feat. As biologists Frederic Mery and James Burns explain, "Behavioural plasticity requires a large behavioural repertoire and selection, using environmental information as a guide, of an adaptive behavioural response. During learning, this selection process is usually performed through trial-and-error, which subjects an animal to a period of suboptimal behaviour. This is the oft-mentioned 'cost of being naïve.'"[87]

The question is whether inherent variation and plasticity are sufficient to explain the origin of complex programmed behaviors. Snell-Rood describes how behavioral plasticity can have significant costs "in time, energy and exposure" because of the trial-and-error selection processes.[88] These issues present a problem for a neo-Darwinian evolution of adaptable behavioral plasticity. Where such plasticity does occur it appears to be specifically engineered to adapt to variable environments, and the behaviors are programmed, as is the case of social insects. Trial-and-error is not a reasonable explanation for the evolution of such behaviors.

Invertebrate Learning

NONE OF the above is meant to suggest that social insects do not learn. The focus of this book is on complex programmed behaviors, so most of the discussion has been on the programmed or innate aspects of the behaviors being explored. But in many cases, the creature succeeds through a combination of programmed and learned behavior. "Learning is probably a universal property of insects," writes animal behavior specialist Reuven Dukas, "which rely on learning for all major life functions."[89]

Honey bees have demonstrated notable learning abilities. This is despite having extremely small brains, consisting of 950,000 neurons and a volume less than one cubic millimeter.[90] The lifeline for bees is nectar, which can only be obtained from certain kinds of flowers. While it may seem a simple task to identify the correct flowers, it isn't. There is a great deal of variability in flowers in terms of nectar volume and concentration, flower depth, size, architecture, and spacing.[91] Tautz identifies the following information that a bee must be able to obtain: recognize the flowers as flowers, distinguish among various kinds of flowers, "recognize the state of the flower" and "know how to work the flower effectively with legs and mouthparts," establish "the geographic location of the flower in the landscape," and "determine the daily time window in which various flowers produce the most nectar."[92]

A well-developed sense of smell and color vision are crucial to such efforts, but while these specialized senses are necessary, they are far from sufficient. How do micro-brained honey bees navigate all these complexities to decide where and on what flowers to forage? Much of this must be learned, and the capacity to do so very quickly appears to be programmed. For example, a bee can learn a specific odor with a single experience and subsequently distinguish it from other odors with 90 percent certainty.[93]

Bees learn the flower characteristics (location, shape, color, and smell) by associating them with the nectar produced by the flower.[94] Experiments have shown that bees show a distinct preference for flowers with higher concentrations of nectar.[95] In addition to learning and associating the physical characteristics of flowers, bees also can learn the optimal time of day to forage at specific locations. This is essential because different species of flowers produce nectar only during certain times of the day. Bees can learn these times and remember the information indefinitely.[96]

Recent research has confirmed the ability of bees to integrate information learned about objects through two different senses (vision and touch). The researchers conclude, "Similar to humans and other large-brained animals, insects integrate information from multiple senses into a complete, globally accessible, gestalt perception of the world around them."[97] In addition to all of these learning abilities, there is also learning involved in navigation, as described earlier. Bees, for instance, learn and remember the navigation information associated with the location of foraging sites and nest locations.

One study compared the neural mechanisms involved in associative learning in fruit flies, honey bees, and crickets,[98] and it found that honey bees and crickets have similar mechanisms, while fruit flies have very different mechanisms. One reason this is surprising is that fruit flies and honey bees belong to the same order (Diptera), while crickets are in a different order (Orthoptera). This means common evolutionary pathways do not appear to provide an explanation; it is what evolutionists call "convergence."

An important issue in evaluating the role of learning in animal behavior is how to assess the costs and benefits of learning versus having fixed innate behaviors. At first glance it would appear obvious that learning provides greater adaptability and thus more of a long-term benefit compared to behavior that is not capable of changing. However, there is a significant cost to being able to learn, not the least of which is the necessary cognitive capacity. An example of the challenges associated with learning is the complexity of the information needed to facilitate it. Learning experiments reveal that most learning only occurs when the delays are less than a few seconds.[99] For example, fruit flies cannot learn to associate an odor with an electric shock when they are separated by more than sixty seconds. Animals live in a complex world (their *Umwelt*) where some information is relevant and some irrelevant; therefore they need to be able to choose which environmental cues, if any, predict the consequence of an action. This makes learning highly complex, and it means that learning, as opposed to innate behavior, is quite costly due to the demands of brain organization and functionality.[100]

One theory that purports to explain the origin of some novel behaviors is the Baldwin Effect, originally proposed by American psychologist James Baldwin in 1895. The theory attempts to account for how learning accelerates the evolution of advantageous behaviors. There are two major steps in the process.[101] First, when animals have behavioral plasticity and the capacity to learn, they can develop new behaviors that are adaptive. When this occurs, selection pressure will then favor the evolution of learning ability. This results in the learned behavior spreading throughout a population. Second, natural selection will favor genotypes that code for the behavior without requiring learning. This eventually leads to the evolution of the genes for the behavioral trait in the population.

While the theory is an attractive one, there is very little evidence supporting it. In any event, it is likely not relevant to the development of the complex programmed behaviors that are the subject of this book, particularly in the case of insects. The reason is that the initial appearance of a novel behavior, according to the Baldwin Effect theory, requires an animal with sufficient cognitive ability to develop it through learning, because it is not innate. And as discussed elsewhere in these pages, evidence suggests that tiny-brained insects lack the cognitive capacity to master the sorts of sophisticated behaviors explored in this book without a significant CPB component.

A quick survey of the repertoire of behaviors of various animals indicates that the capacity for learning is usually very limited. In addition, the learning that does occur is specific to the behaviors critical to the animal and is not open-ended. A study by Stefano Ghirlanda, Magnus Enquist, and Johan Lind based on modeling the development of learning found that a large behavioral repertoire exacts a cost by increasing the time required to master functional sequences of behaviors. This conclusion was based on two observations. One is that behavioral repertoires are primarily genetically (or possibly epigenetically) determined, and they often are limited to just a few behaviors. Two, animals usually combine "genetically determined behaviors into learned sequences."[102] And as the authors pointed out, as the number of individual actions increases, the number of combinations of these actions increases exponentially. The problem then is that the probability of learning how to combine a coordinated sequence of a number of actions to produce a useful behavior decreases exponentially. Think of the amount of learning that would be involved in the numerous behaviors of social insects, which explains why much of it is programmed.

An impressive combination of learning and programming can be found in other "simple" invertebrate organisms, such as nematodes (roundworms), including the species *Caenorhabditis elegans*. Their nervous system consists of only 302 neurons and approximately seven thousand synapses, and it is the first animal to have its nervous system completely mapped.[103] It was once thought that the behavior of C. *elegans* was totally programmed, but research has shown otherwise. Their learning includes both short-term and long-term memory.[104] They can integrate and remember experiences based on information obtained from different senses, as described above with bees. And they can learn to associate food with or without different substrates (e.g. salt). As University of British Columbia psychologist Catherine Rankin emphasizes, even the primitive C. *elegans* worm is "exquisitely designed to benefit from its experience."[105]

How can a species whose brain contains only 302 neurons demonstrate such an impressive combination of learned and programmed behaviors? According to an article on C. *elegans* in the journal *Cell*, "It remains largely mysterious as to how the nervous system is functionally organized to generate behaviors."[106] What makes it even more mysterious is that all nematodes appear to have very similar nervous system designs, while exhibiting a wide variety of behaviors. Cambridge University neuroscientist William Schafer lists several proposed explanations, including expressing different receptors or neurotransmitters, substantially reorganizing synaptic connectivity, and "individual neurons adopting distinct functional properties in different species."[107]

While much here remains mysterious, several inferences can be drawn from these findings. One is that the behaviors are not defined solely by genes or by neural architectures in the brain. There are several other possible factors that determine CPBs. Second is that even relatively simple behaviors are controlled by complex neural systems, even in animals such as *C. elegans* with only 302 neurons. Taken together, these many factors demonstrate the design complexity underlying the behaviors even in animals marked by extreme anatomical simplicity.

Darwinian Evolution and Social Behavior

THE COMPLETE programmed behaviors involved in social organization are controlled by multiple algorithms. While there is not an overriding blueprint that governs the social behaviors, there is an integrated set of rules associated with each behavior. Hölldobler and Wilson describe the algorithms as being implemented through a combination of "decision rules," and note that "a complete sequence of decision points that produces a caste, product, or full behavioral response is called an algorithm."[108]

They define two classes of algorithms. One class operates when a colony member is either an egg or larvae, and determines whether it will become a worker or a breeder. In some species there is also a second decision related to whether the individual becomes a minor or major worker. Some major workers become soldiers in an army (i.e., army ants). When the individual becomes an adult, its behavior is controlled by the second class of algorithms, performing labor appropriate to its caste and adult age. The insect responds to the stimuli its sensory and nervous systems are programmed to respond to.[109] Most of the decision points involve simple selections between two alternative behaviors.

However, in a further complication, social insects are also programmed to learn and adjust their behavior as circumstances change. For example, an ant repairing damage to a nest wall who comes across a misplaced larva will switch to a different behavior, pick up the larva and return it to its proper place in the brood chamber. In this case, the latter behavior (larva care) is programmed to have a higher priority over the former behavior (wall repair). That is a programmed decision.

Hölldobler and Wilson ask an important question: "How does a superorganism arise from the combined operation of tiny and short-lived minds?"[110] After all, "Nothing in the brain of a worker ant represents a blueprint of the social order. There is no overseer or 'brain caste' who carries such a master plan in its head."[111] Hölldobler and Wilson theorize that "natural selection at the colony level creates algorithms that maximize efficient order. Genes prescribe algorithms, which guide colony members by means of the nuances of sensory thresholds, context, and innate flexibility in a manner that draws the appropriate response from the colony as a whole. It is the integration of these modules of individual behavior that determines the fate of the colonies and hence of the genes that prescribed their construction."[112]

But the idea lacks a sound evidential basis. Their basic premise is that even simple rules with binary (yes/no) decisions can create an extremely complex set of algorithms, given a sufficient number of rules, and that random variations could add in favorable binary decisions bit by bit until the full symphony of programmed behaviors among different colony members was in place.[113] But while each decision rule may be relatively simple, any new decision rule must be fully integrated into the entire and increasingly complex body of rules and algorithms.

And there's the sticking point. As evolutionary biologists Joan Strassmann and David Queller note, "No individual is doing anything that by itself would be very useful; instead, each is performing a role in a process that only makes sense in terms of increasing colony function."[114] The problem is that a large number of combinations of rules are subject to random variation and mutation. In many instances, each change to a decision rule is just as likely, if not more likely, to have an unfavorable outcome as a favorable one. Further, while each decision rule may be relatively simple, it still must be fully integrated with the entire body of rules and algorithms. They must all be coordinated and invoked under the appropriate conditions and in the appropriate order. Any random combination of rules results in chaos. And indeed, if any single rule were to go drastically wrong, the result could be the death of the colony.

One puzzle concerning the origin of social behavior is the fact that it is present in only a few groups of insects. Eusociality exists in several cases among Hymenoptera (including ants, bees and wasps), and once among termites.[115] One phylogenetic study supported the hypothesis that these three groups share a common ancestor.[116] However, the evidence indicates the theoretical common ancestor shared some traits with what are considered its descendants, but was not eusocial. The general consensus is that "ants (family Formicidae), bees (superfamily Apoidea), and wasps (families Vespidae and Crabronidae) have all evolved social behavior independently."[117]

There are several competing theories about the evolution of insect social behavior. Following is a review of them.

Behavior Toolkit

SOME RESEARCHERS believe there is a relatively straightforward process to transition from solitary insects to the division of labor found among eusocial animals. The idea is known as the "groundplan hypothesis." According to this hypothesis, what are regarded as simple changes in the regulation of solitary behaviors result in the complex set of behaviors associated with sociality.[118] What's the trick? According to the groundplan hypothesis, the complex behaviors were already present in the non-social ancestor. Hölldobler and Wilson, who are advocates of the groundplan hypothesis, explain: "The division of labor appears to be the result of a preexisting behavioral groundplan in which solitary individuals tend to move from one job to another after the first is completed. In eusocial species, the algorithm is transferred to the avoidance of a job already

being filled by another nestmate. It is evident that progressively provi-
sioning bees and wasps are already 'spring-loaded' for a rapid shift to
eusociality once ecological factors favor the change."[119]

But just divvying up the roles isn't enough to produce a functioning
colony. These behaviors include reproduction, brood care, nest construction, defense, and foraging. Taken together, they are necessary but not
sufficient for colony survival. The systems of interdependent role players
require social pathways regulated by a communication system through
multiple pheromones and other signals, and links between the sensory
system to the various social behavioral modules. Evidence for this comes
from studies that have found that when honey bee workers were exposed
to pheromones, it led to changes in their brain gene expression which, in
turn, altered their behavior.[120]

There are other problems with the groundplan hypothesis. As Johnson and Linksvayer note, "An advanced social insect colony has a task
repertoire approximately twice that of a solitary insect.... Many highly
derived caste-specific traits have no correlates in solitary wasps or bees.
The advanced brood care behaviors of honey bees, for instance, provide
a clear example of a... eusocial behavior without an obvious antecedent
in solitary bees."[121] Such behaviors cannot be borrowed from the solitary
ancestor because the solitary ancestor did not possess the behavior.

Also, "colony-level regulatory processes control the expression of
caste-generating gene sets. In short, whether a larva develops into a
queen or a worker depends on social environmental factors—such as
nutrition—that are regulated by the colony."[122] Or as they put it more
generally, "Caste-specific expression of groundplan genes is controlled
by social feedback loops—that is, social physiology."[123] (See Figure 4.1.)
So, where did these colony-level regulatory processes that control the
expression of caste gene sets come from? Not from the solitary insect
ancestor.

If a computer programmer were tasked with writing the software
for such feedback loops in, say, a set of interactive robots, the task would

require exquisitely complex programming of the artificial intelligence. So, how did these sophisticated feedback loops arise in insects?

Novel Genes and Social Behavior

A SIGNIFICANT question about the origin of the genetics underlying insect social behavior concerns the number of novel or modified genes required for these behaviors. Research indicates a relationship between novel traits and so-called "orphan genes." An orphan gene is a gene unique to that particular species.[124] The term orphan is a bit of a misnomer, as the evidence indicates they do not have a lineage history.

These orphan genes are evolutionary curiosities, but far from unusual, as emphasized in this summary from *Trends in Genetics*: "Every eukaryotic genome contains 10–20% of genes without any significant sequence similarity to genes of other species; these are classified as 'orphans' or 'taxonomically-restricted genes' (TRGs).... such genes have arisen in the genomes of every group of organisms studied so far."[125] This ubiquity suggests that they may code for novel traits. "It is becoming increasingly clear that they have contributed substantially to the evolution of organisms and evolutionary innovations," write geneticists Diethard Tautz and Tomaslav Domazet-Loso.[126]

But their widespread presence poses a challenge to Darwinian evolution, with its commitment to common descent and a process of very gradual diversification. On the Darwinian model we aren't primed to expect so much novelty with no apparent evolutionary relatives. And yet despite the mystery hovering around these orphan genes, "few orphan genes have been studied in much functional detail so far," Tautz and Domazet-Loso could state even as late as 2011.[127] Or as the *Trends in Genetics* paper noted around the same time, "Oddly enough, most attention at present has been paid to genes which are shared and highly conserved throughout evolution, and not to those which are unique or lineage-specific."[128] It's as if evolutionists wanted to pay attention to the part of the picture that makes sense to them on evolutionary grounds, and were tempted to give short shrift to the parts that don't.

Because orphan genes have no apparent evolutionary antecedents, their origin has sometimes been called "*de novo*"—a vague term to encapsulate their seemingly spontaneous functional appearance in the genome from non-protein-coding DNA. German molecular biologist Henrik Kaessmann writes that a *de novo* origin of protein-coding genes has long been viewed as very unlikely but that "recent work has uncovered a number of new protein-coding genes that apparently arose from previously noncoding (and nonrepetitive) DNA sequences… It is now firmly established that new genes have indeed been major contributors to the origin of adaptive evolutionary novelties."[129]

From a design-based paradigm, the notion of a *de novo* origin of a complex feature is not unreasonable since an intelligent agent can rapidly insert information into a system. But a neo-Darwinian paradigm struggles to account for such data, because unguided processes make the *de novo* origin of protein-coding genes highly improbable.

As noted above, Hölldobler and Wilson, in keeping with their groundplan hypothesis, assert that only minor genetic changes are required for a species to transition to eusociality. "It seems to follow as an overarching principle that a change on the order of the final step to eusociality can occur with the substitution of only one or a small set of alleles," they write.[130] However, recent research suggests otherwise.

The rapidly developing field of genomics focuses on the structure and function of genomes. Its sub-discipline, sociogenomics, relates genomics to social behavior. The genomes of many animal species have now been decoded, including those of several social insects, enabling researchers to determine the types of genetic changes associated with social behavior. The relationship is apparently far from simple. The results from several studies indicate that significant genetic changes are associated with social behavior.

Johnson and Linksvayer's analysis of the groundplan hypothesis in light of some of these findings and focused on the Hymenoptera insect order concluded that "the communication networks that characterize social physiology have no direct antecedents in the solitary Hymenoptera" and that "the evolution of eusociality involves a complex mix of changes in gene expression, genes with modified functions, and novel genes."[131]

A study that compared the genes among several groups of bees with varying degrees of sociality found a common set of 212 genes that exhibited what the researchers regarded as "accelerated evolution."[132] In addition, there were 173 genes that showed accelerated evolution among the highly eusocial groups.[133] The term "accelerated evolution" refers to changes in the genome that are far greater than would be expected due to normal mutation rates. But labeling the genetic changes in this way isn't a causal explanation; rather it simply begs the question of whether any blind evolutionary process could generate such changes so quickly.

Another study of the genomes of ten bee species assessed the genetic changes associated with social behavior.[134] The study examined the genomes of solitary bees, bees that exhibit simple eusociality, and those that exhibit complex eusociality. The researchers reported the following: 1) As they examined the transition from the solitary bees to the increasingly eusocial bees, they found an increased capacity for gene regulation in the eusocial bees, including "rapid evolution" of the genes involved in coordinating gene regulation. There were 162 genes exhibiting what the researchers described as "accelerated evolution." 2) The bees exhibiting increased complex eusociality had 109 genes significantly enriched for functions tied to protein transport and neurogenesis. 3) There was rapid evolution of signal transduction pathways. 4) The genetic changes differed among the various lineages that developed eusociality, indicating there was not a single pathway, though the various pathways likely all involved an increase in the complexity of gene networks.[135]

Another hypothesis about the evolution of social behavior postulates a high level of genetic recombination in social insects. Genetic recombination occurs due to the phenomena of crossing over of genes on chromosomes. This produces new gene combinations on the same chromosome. According to this hypothesis, complex eusocial colony behavior could have developed due to an increased rate of recombination. However, a study comparing the genomes of the eusocial honey bee

(*Apis mellifera*) and the solitary alfalfa leafcutter bee (*Megachile rotundata*) found "no evidence that recombination has modulated the efficacy of selection among genes during bee evolution, which does not support the hypothesis that high recombination rates facilitated positive selection for new functions in social insects."[136]

A study of the highly eusocial Asian honey bee (*Apis cerana*) genome found 2,182 unique genes out of a genome consisting of 10,651 genes—about 20 percent of the total genome.[137] In addition to these unique genes not being shared with other non-social insects, the closely related western honey bee (*Apis mellifera*) also does not share commonality with these genes. That is surprising, since it is believed the two species diverged from a common ancestor only one million to two million years ago.[138] In addition, the Asian honey bee has a number of different behavioral traits, including group-level defense, grooming, and hygienic behaviors.[139] These findings are another indication of the number of novel or orphan genes associated with complex social behavior.

Similar research was conducted on paper wasps (*Polistes canadensis*), which exhibit a limited amount of social behavior, especially when compared to many ant species. The results found approximately 2,500 genes associated with social behavior.[140] Another study of the expression of genes associated with the social behavior of one ant species (*Temnothorax longispinosus*) found over five thousand genes that were differentially expressed, meaning that the amount and timing of the gene products (usually proteins) varied.[141]

According to Seirian Sumner, a British entomologist and behavioral ecologist, the findings in these genome studies indicate that novel genes are "an important source of phenotypic innovation across the animal kingdom, including caste-biased genes in social insects. We now understand that there is substantial gene birth and evolution going on among genes that are up-regulated in the worker, but not the queen, caste. The evidence for this now comes from the three main clades of social Hymenoptera (bees, wasps and ants)."[142]Again, it is an assertion that the novel genes are the result of a neo-Darwinian evolutionary process, but for which there is little to no evidence.

Another study compared the genetic changes in seven species of ants, honey bees, and several solitary insects. This study resulted in several significant findings. Each lineage of ants contains about four thousand novel genes, compared to solitary insects.[143] However, there are only sixty-four genes that are shared by all seven species of ants, undercutting the notion of a basic toolkit. The study authors conclude, "These results suggest that a broad 'social toolkit' of conserved *de novo* protein-coding genes is not a requirement for eusociality."[144]

A related area of research concerns the non-coding regions of the genome.[145] These areas used to be referred to as "junk DNA" because they were thought to lack function, an assumption encouraged by the neo-Darwinian paradigm. But a recent study of bee genomes found otherwise. "Novel non-coding aligned regions (NCARs) recruitment is associated with the emergence of obligate eusociality in the corbiculate bees (*Apis, Bombus, Melipona* and *Eufriesea*)," a team of researchers concluded. "A total of 1476 NCARs are shared uniquely among these species and are enriched for gene functions associated with cell and nervous system development."[146] The finding seems to indicate that "the origin of eusociality in [corbiculate bees] was accompanied by an increased regulatory capacity provided by these NCARs."[147] These findings align with analysis of vertebrate genomes, suggesting that the origin of novel phenotypes is correlated with changes in conserved, non-coding elements associated with gene regulation.[148] The research provides evidence to support a hypothesis that one of the potential origins of complex novel behavior is due to novel changes in the non-protein-coding regions of the genome.

Epigenetics

As MENTIONED previously, due to the failure of the neo-Darwinian mechanism to explain the origin of novel traits and other phenomena, various additional evolutionary mechanisms have been proposed under the "extended evolutionary synthesis." One of these focuses on what is known as *epigenetics*. In 1942, the renowned British biologist C. H. Waddington introduced the word "epigenetics" to mean the study of "the processes involved in the mechanism by which the genes of the genotype bring about phenotypic effects."[149]

Epigenetic processes can cause changes in when or where a gene is expressed, but not in a gene's DNA sequence. This is a challenge to the fundamentals of the modern evolutionary synthesis because it acknowledges information affecting morphology and function outside the DNA sequence. According to neo-Darwinism, the crucial evolutionary action takes place via random mutations in the DNA. But if important aspects of morphology are affected by biological information outside the DNA sequence, then how could the genetic mutations of the neo-Darwinian mechanism suffice? The key question for those attracted to the emerging field of epigenetics but also committed to evolutionary theory was how to extend the modern synthesis to include insights from epigenetics.

Epigenetics has been the subject of significant research in many types of animals, including humans. The results of several studies suggest that "some behavioral variation, and phenotypic plasticity in general, is mediated by *epigenetic* mechanisms, molecular-level processes... that modify gene expression but do not change DNA sequence and may lead to heritable change in phenotype."[150]

Epigenetics also has shed fresh light on embryology. Many biologists assumed that an embryo's development is completely controlled by genes, an assumption dictated by the "central dogma of biology" that DNA sequences are the sole repository of information needed for the development of organisms. However, evidence suggests there is ontogenetic information not specified by DNA. In a paper surveying the evidence, Jonathan Wells concludes, "The idea that embryo development is controlled by a genetic program is inconsistent with the biological evidence. Embryo development requires far more ontogenetic information than is carried by DNA sequences. Thus Neo-Darwinism is false."[151] Based on the evidence that there are other sources of information that control development, Wells observes that "an adequate theory of evolution would have to explain how various information sources in the organism (including its DNA, membrane patterns, and cytoskeletal structure) change in a coordinated fashion to produce new species, organs, and body plans."[152]

There are numerous examples of the role of epigenetics in the development of animal behavior. Research has shown that "stressful experiences from the parents are passed to future generations by epigenetic changes in stress-related genes."[153] The research showed that maternal hormones can significantly impact bird chicks. Offspring from laying hens that were stressed and thus released more stress hormones were more fearful and less competitive. Similar epigenetic effects have been observed in other species.

The behaviors expressed are already potentially present in the phenotype, as in the case of social insect caste determination. In the case of social insects there is evidence that epigenetic processes that regulate gene expression could provide the major mechanism for the regulation of neuronal memory and other properties of the brain.[154] Despite this evidence biologist Hua Yan and his co-authors conclude, "The major questions that drive the fields of behavioural epigenetics and sociobiology remain unanswered. What are the principle aspects of a genetic and epigenetic architecture that direct the production of distinct morphological and behavioural castes? What is the extent to which epigenetic modifications drive social context-dependent behavioural plasticity in brains?"[155]

The studies described above show that epigenetic mechanisms affect the development of castes and expressions of insect social behavior. However, such mechanisms are insufficient to have evolved social insect CPBs in the first place. The caste characteristics and behaviors expressed are already potentially present in the phenotype. Additionally, in many cases the epigenetic mechanisms are themselves quite complex. They appear to be one of the controlling factors in the execution of CPBs. Therefore, their origin also requires an explanation, and their sophistication poses a challenge for any unguided evolutionary scenario.

Social Behavior Altruism

Another issue with origin of social behavior is explaining the origin of altruism in the context of neo-Darwinian evolution. Darwin recognized the difficulty in explaining the evolutionary origin of social behavior, specifically the castes that do not reproduce, when he wrote about "one special difficulty, which at first appeared to me insuperable, and actually fatal to my whole theory." He elaborated thus:

> I allude to the neuters or sterile females in insect-communities: for these neuters often differ widely in instinct and in structure from both the males and fertile females, and yet, from being sterile, they cannot propagate their kind... But with the working ant we have an insect differing greatly from its parents, yet absolutely sterile; so that it could never have transmitted successively acquired modifications of structure or instinct to its progeny. It may well be asked how is it possible to reconcile this case with the theory of natural selection?[156]

Under the Darwinian paradigm and the "selfish gene" concept view espoused by Dawkins and others, the Darwinian process rewards the individuals who survive and reproduce, not those who are unable to reproduce and help others to do so.[157] So, why then do we find striking cases of altruistic behavior in the animal kingdom, as in social insects?

The theories of inclusive fitness and kin selection are attempts to harmonize such cases with Darwinian evolutionary theory. Inclusive fitness, originally developed by evolutionary theorist William Hamilton, has been cited by many as the theory that explains the evolution of social behavior, in particular the altruistic traits that are common in social societies. Hamilton's theory is intended to explain why some social workers are sterile and altruistic.[158] The basic idea is that animals will act to benefit closely related individuals because they share a significant percentage of their genes.

While the debate about inclusive fitness is certainly interesting, the resolution of it one way or the other does not directly affect the thesis of this book. The reason is that an explanation of the origin of the social behaviors and the algorithms that control them has been left out of the debate. Inclusive fitness theory and other Darwinian mechanisms simply assume the existence of these behaviors and algorithms. It does not explain their origin, other than relying on the standard Darwinian assertion of selection pressure. Inclusive fitness theory and group selection are only able to explain how the behaviors are adaptive and maintained within a population.

Before recent genome research, it was assumed that only relatively minor genetic changes were required to transform insects from solitary to social. That assumption has been decisively refuted, based on the findings of the studies cited above. The development of social behavior is not simply a matter of maintaining a basic toolkit of genes for generic behaviors and the evolution of a small number of new genes. It is now known that the transition to social behavior requires hundreds or thousands of modified or novel genes and their expression through epigenetic mechanisms.

A problem in the studies of genetic changes associated with eusocial behavior is that they do not appear to account for the origin of the algorithms that control the behaviors at the colony level. One possible reason is that the algorithms do not reside within the protein-coding genes, as there is evidence of changes in non-protein-coding DNA associated with social behavior. There is also some evidence that aspects of social behavior are controlled epigenetically.

Regardless of the exact mechanism, the complexity of the algorithms presents a major challenge to a Darwinian explanation, and indeed to any explanation committed to wholly blind evolutionary processes, because numerous new systems of behavioral control and logic must appear before these behaviors can provide an advantage. As the research has indicated, hundreds or possibly thousands of novel or modified genes have been found to be associated with social behaviors. Blind evolutionary processes are poor candidates for causing such a large number of functional mutations and for coding such complex systems. Researchers are more than justified in casting about for an alternative hypothesis that identifies a cause with the demonstrated ability to produce the effects in question.

5. Insect Architecture

"What's miraculous about a spider's web?" said Mrs. Arable. "I don't see why you say a web is a miracle—it's just a web."

"Ever try to spin one?" asked Dr. Dorian.[1]

— E. B. White, *Charlotte's Web*

THERE ARE STRUCTURES BUILT BY ANIMALS THAT HUMAN ARCHItects marvel at for their skillful engineering and artistry. The fact that these structures are produced by animals, many with tiny brains and minimal intelligence, makes it all the more remarkable.

Architecture is employed by a range of creatures across the animal kingdom. Birds and insects build nests. So too do some spiders, crustaceans, fish, reptiles, and mammals. The primary function of these home-building efforts is security, mainly related to raising offspring. In addition, this behavior promotes continuity of generations and more socially complex lifestyles.[2]

An important question about architectural behavior is whether it is innate or learned. James and Carol Gould explain that given the enormous complexity of bird nests and beaver dams, early authorities assumed that elaborate construction behavior must be learned. However, naturalists disagreed, since they observed nests being built by "inexperienced first-timers and saw that, despite imperfections, they had most of the characteristics seen in the constructions of mature adults."[3] In other words, their behavior appeared to be largely innate and programmed.

Further evidence that some animal architecture is innate comes from the fact that some insects can build elaborate structures. When we rank animals according to their ability to build sophisticated structures, it turns out that social insects such as ants and termites are the closest to talented human architects, far outstripping the best efforts of even the most clever mammals outside of humans.[4] These insects, with their miniscule brains, simply lack the gray matter to master such architectural marvels through the sort of intellectual growth and learning that humans undergo on the way to becoming successful architects.

The fact that these insects almost immediately master their architectural skills while still young (much as birds and beavers do) also strongly suggests that we are witnessing a complex programmed behavior.

Bee Hives

DARWIN EXPRESSED his admiration for the design of honey bee combs. He describes how it might seem inconceivable that bees are capable of such an engineering feat, writing, "He must be a dull man who can examine the exquisite structure of a comb, so beautifully adapted to its end, without enthusiastic admiration." He continues:

> We hear from mathematicians that bees have practically solved a recondite problem, and have made their cells of the proper shape to hold the greatest possible amount of honey, with the least possible consumption of precious wax in their construction. It has been remarked that a skillful workman, with fitting tools and measures, would find it very difficult to make cells of wax of the true form, though this is effected by a crowd of bees working in a dark hive. Granting whatever instincts you please, it seems at first quite inconceivable how they can make all the necessary angles and planes, or even perceive when they are correctly made.[5]

According to Sophie Cardinal and Bryan Danforth, entomologists at Carleton and Cornell universities, respectively, research suggests that the oldest form of hymenopteran did not engage in complex nest construction or eusocial behavior.[6] The Goulds theorize that complex hive construction and social behavior developed concurrently: "The first step toward sociality was architectural: some species began building protective structures for their larvae such as burrows."[7] Interestingly, sociality in bees is said to have developed independently several times, "and each time novel strategies of communal nest building, social hierarchy, and communication had to be invented."[8] The reason communal nest build-

ing appears to have to been invented independently these several times is
that this behavior does not exist in non-social bees.

Honey bee hives are impressive wherever you find them, but especially so in colder climates due to the additional features these hives require to allow the bees to survive the winters. Insect behavior specialist
Bernd Heinrich describes the essential functional characteristics of the
hive in colder climates:

- The nest site must protect the bees from deep frosts.
- It must be spacious to hold large amounts of honey and pollen in
  the winter.
- The young must be fed and kept warm in order to grow. This
  means the home must be heated, requiring a constant use of fuel.
- Honey bee homes must be large enough to accommodate tens
  of thousands of bees and store enough honey to last through the
  winter.
- Because they contain large amounts of rich honey and brood
  nurseries that tempt predators, they must also be heavily
  defended.[9]

These and other functional requirements explain much of the design of honey bee hives. There are several notable aspects of comb architecture. The cells are built out horizontally, attached back to back in a
double layer. Each three-sided half of a given hexagonal cell also provides
walls for three adjoining hexagonal cells, one side for each, which is a
"structural trick discovered millions of years later by human engineers."[10]
An average colony will have several sheets of parallel comb with about
five thousand cells in each. The comb sheets are made of beeswax, with
the strength such that one ounce of wax honeycomb can hold about two
pounds of honey, pollen, larvae, and pupae.[11] Honey bees also manufacture and apply a waterproof lining to their hives. Some species create
the material by collecting plant resins and tree sap, which they then mix
with pulp or mud to create a substance called *batumen*.[12]

It was long believed that beeswax was gathered from flowers, along with nectar and pollen.[13] Eventually it was determined that it is synthesized by the bees themselves. This makes bees one of the few animals able to manufacture nest material from their own bodies. As behavioral biologist Jürgen Tautz explains, the wax is extruded from "eight groups of glands arranged in pairs" on the abdomen, and after "being extruded onto the body surfaces of the bee, it hardens into small, paper-thin scales."[14]

The amount of work that goes into building a comb is impressive. To construct an average comb from scratch requires the nutritional energy provided by about sixteen pounds of honey in order for the industrious bees to produce about two and half pounds of wax, which is used to construct about 100,000 cells.[15] Heinrich comments that "the precise regularity of honeycomb has been a source of wonder, if not astonishment. There has been a debate about how the bees create the shapes of the cells. After 350 scientific papers about research on honey bee comb… we still don't have the full answer to how bees make honeycomb."[16]

One tool in the honey bee's bag of tricks for building highly regular bee nests is the use of the earth's magnetic field. While the use of the magnetic field is common in navigation, its application to nest building is unique. The Nobel Prize-winning German-Austrian ethologist Karl von Frisch found that in certain conditions bees orient the comb to magnetic north,[17] enabling them to have a common orientation and framework. Without this, coordinating the efforts of individual bees would be much more difficult. Further studies have found that this is just one of several methods the honey bees use for orientation.[18]

Another important element of bee hives is the characteristics of honey. It is a super-concentrated source of energy whose high density, acidic nature, antimicrobial enzymes, and relatively low water content mean that molds and bacteria cannot grow in it.[19] This connects to another characteristic of honey essential to colony survival: it can be stored, an innovation that helps shield bees from environmental fluctuations in available nectar.[20]

Development of Honey Bee Combs

CHARLES DARWIN waxed eloquent about the construction of honey bee combs but did not waver in his view that the origin of this skill lay within the reach of his theory of evolution: "Thus, as I believe, the most wonderful of all known instincts, that of the hive-bee, can be explained by natural selection having taken advantage of numerous, successive, slight modifications of simpler instincts; natural selection having by slow degrees, more and more perfectly, led the bees to sweep equal spheres at a given distance from each other in a double layer, and to build up and excavate the wax along the planes of intersection."[21]

However, Darwin seems to ignore that comb construction involves several essential bee behaviors that are coordinated and necessary to successfully produce a comb. These individual behaviors include selecting and preparing the tree cavity, constructing the three-dimensional shape of the hive, selecting and synthesizing materials, maintaining a critical temperature, and coordinating construction activities among the bees. All of these are critical elements and are interdependent, meaning they arguably work as a kind of irreducibly complex system of behavioral systems. Remove one or two and you soon get a dead bee colony. If it's not quite all or nothing, it's close. This poses a severe challenge to gradualistic evolution.

Is it possible that comb construction behavior has evolved to some degree over time? That is not only possible but likely. It is clear that each of the individual behaviors is subject to natural selection, which could result in improvements to comb construction, allowing the hives to better meet changing environmental challenges. However, that does not provide an explanation for the origin of the individual behaviors, the algorithms that control the behaviors, and the complexity of the social colony and coordination among the bees. These considerations pose significant problems for Darwin's idea that such architectural capacities arose through natural selection working on a long series of slight, random modifications.

Wasp Nests

ORGAN-PIPE WASPS construct nests that have the appearance of organ pipes, hence their name. The nests consist of a series of tubes made of mud (the wasps are also called mud daubers), in which eggs are laid and allowed to develop into larvae. To construct the organ-pipe nests, the female wasp gathers mud with her mandible, forms it into a ball, returns to the nest, mixes the mud with saliva, then spreads out the mud balls into strips using her forehead. According to the Goulds, this building behavior is "so stereotyped that it must be based on a set of coordinated programs." It's true that, as the Goulds also note, "later pipes show definite improvement over earlier ones,"[22] so there is some learning involved, but the basic behavior is innate and programmed. The Goulds further note that experiments show that the wasps do not have any cognitive picture of what the finished nest should look like. Instead, they appear to exhibit a series of consecutive motor programs in completing each step of the building process. What is also remarkable: the organ-pipe wasps manage these architectural marvels despite extremely poor vision, roughly 20/2,000.[23]

One of the major advantages of nest building is the control of the internal climate. This allows species to survive in a larger variety of climates. This is illustrated in the paper nests built by some species of paper wasps. Paper wasp nests are quite complex internally, involving multiple layers and paths to traverse them. Climate control involves both heating and cooling. Heating is achieved by a group of adult wasps forming a cluster and pulsating their wing muscles, thus generating heat. Cooling is achieved when water is brought into the nest, which then evaporates, causing a cooling effect.[24] The paper nests can also include insulation, which the wasps construct out of a series of concentric spheres around the comb. Temperatures are kept to within three degrees of 86°F.

Weaver Ant Nests

MOST ANT species spend most of their lives underground, using tunnels for their living quarters. However, a small number of species live in trees.

One such species is called the weaver ant (genus *Oecophylla*), so-called because these ants construct nests (typically the size of a soccer ball) by weaving leaves together. Weaver ant colonies can reach populations of a half million ants, consisting of a hundred nests spread over numerous trees.[25]

The remarkable thing about weaver ant nests is not so much the structure as the method employed to construct them. One unique element is that the leaves used in the nest remain attached to the tree and thus continue to photosynthesize, producing oxygen. But the most remarkable aspect is the teamwork involved by the community of ants.

As shown in Figure 5.1, the ants sew leaves together by joining the edges of leaves. They work meticulously in joining the leaves by aligning the two edges and then drawing them together. If the edges of the leaves are jagged, they trim them to ensure the two edges are even. In instances where the edges do not reach, the ants form an ant "chain gang" where several ants work together to bridge the gap. The leaf edges are glued together using the silk extruded by ant larvae—silk that would normally be used to construct an individual cocoon. The larvae thus modify their behavior, cooperating with the worker ants in the nest-building effort. Put another way, during this process the worker ants use the larvae as tools and control them in forcing the silk to be extruded precisely along the leaf edges. Hence, weaver ants can be classified as tool users!

Another interesting element of this nest-building behavior concerns the initial joining of the leaf edges. The ants work independently or in small groups as they pull on the edges. When one group starts having success, the other nearby ants stop their effort and join the group having the most success. The Goulds comment: "This instance of cooperative building depends on the ability of independent insects, each with its own agenda, to recognize a partially completed project, one further along than their own, and join in."[26] Fascinating videos of weaver ant nest construction can be found on the internet.[27]

Bert Hölldobler and E. O. Wilson ask, "How could such extreme behavior have evolved in the first place?"[28] They point out that it appar-

Figure 5.1. Weaver Ant Nest Construction

ently has evolved independently in four different groups of ants, since the four genera of ants that exhibit this behavior are distinct from each other. "We find it surprising that communal nest-weaving has arisen only four or so times during the one hundred million years of ant evolution," they write.[29] But it is not clear why they are surprised that such extreme behavior arose only in four groups of ants. They seem to assume that such a trait arises easily, and assume this without addressing the many complexities required to evolve such a multifaceted trait in a stepwise Darwinian fashion. Taking all those complexities into account, it seems much more surprising that a blind evolutionary process would create the behavior even once. That it originated four separate times only deepens the mystery.

Hölldobler and Wilson speculate as to how this behavior could have evolved. They observe that in some species the behavior of the larvae is only slightly altered from what it is normally. Therefore, "It is easy to imagine such a change occurring with the alteration of a single gene affecting the weaving program. Thus, starting the evolution of a population toward communal weaving does not require a giant or otherwise improbable step."[30] Hölldobler and Wilson also describe how the weaving behavior is developed to varying degrees, with some being more advanced than others. They flag this as a potential problem for evolution, asking "why so many intermediate species possess what appear to be 'imperfect' or at least mechanically less efficient adaptation?"[31] They proceed to speculate about possible explanations for the diversity based on blind evolution. The more relevant question in the context of this discussion is, How can Darwinian evolutionary processes explain the origin of this behavior at all? It cannot be attributed to the cognitive ability of the ants, which is extremely limited. Clearly this is a complex programmed behavior.

One problem with Hölldobler and Wilson's simple explanation is that it ignores the worker ants' use of the larvae in constructing the nest. This behavior differs much more radically from the norm, and as with all complex programmed behaviors, these ants' behavior cannot be under the control of a single gene. Rather, there must be a complex suite of genes involved, so this behavior is very unlikely to have evolved through random mutation and selection. The reason, recall, is that both observational evidence and conclusions drawn from probability analysis argue against random mutations being able to generate a suite of coordinated genes, with or without natural selection, and yet it appears that a suite of genes is precisely what is needed to create a functional advantage in weaver ant architectural behavior.

Termite Nests

SOME TERMITES form large colonies and farm fungi that digest cellulose, producing the chemical byproducts the termites feed on. Termite colonies can have as many as a million individuals. As you would imagine, such colonies require enormous nests. Many species build their nests underground, which is not surprising as termites are blind and prefer living in the dark. The typical nest includes a royal chamber, nurseries, gardens, waste dumps, a well, and a ventilation system.[32] Some species do build their homes above ground, mounds as large as twenty-three feet high or more (see Figure 5.2). Making this even more impressive is the fact that termites are one-quarter of an inch long, meaning that such mounds are the equivalent of humans constructing a building more than six thousand feet high.

One species of termites (*Amitermes meridionalis*) in Australia constructs nests that are long and skinny. The narrow sides face north and south, and the broad sides, east and west—hence the name of the insects, compass termites. This design minimizes temperature increase due to the sun, as the rising and setting sun are less intense, while at noon when the sun is most intense, only the narrow side of the nest faces the sun.

Figure 5.2. Cathedral Termite (*Nasutiteremes triodiae*) Mound

Many termite nests also incorporate ventilation systems, which not only cool but also remove carbon dioxide. For example, the termite species that farms fungus in Africa (*Macrotermes natalensis*) engineers nests such that the interior temperature is maintained within about one degree of 30°C, and the carbon dioxide concentration between 2.6 and 2.8 percent.[33] Similar measurements of the mounds of another African termite species (*Macrotermes michaelseni*) found the carbon dioxide concentration was maintained between 4 and 6 percent, and the temperatures within five degrees of a nominal 30°C.[34] And this was in an environment where the daily external air temperature varied by 15 to 20°C.

Recently, engineering analyses determined the design methods employed to achieve such stability. One such study found that the nests of an African termite species (*Trinervitermes germinatus*) have outer walls containing both small and large pores that enhance the efficiency of the ventilation system.[35] The permeability of the larger is two orders of magnitude greater than the smaller pores. The larger pores also allow for efficient drainage of water following rainy periods. More impressive, though, is the fact that the termites actively manage the pores by opening or closing them in response to the ambient level of carbon dioxide. This appears to be another programmed behavior.

Termites also can adapt nest construction to local conditions. For example, as a 2019 article reported, the species *Macrotermes bellicosus* builds two different types of nests depending upon the local environment: "thin-walled, cathedral-shaped mounds in open savannas, but thick-walled, dome-shaped mounds in forested areas."[36] One unique design is constructed by the termite genus *Cubitermes* that dwells in rain forests. Their nests include umbrella-like roofs with eaves, which keep water out of the nest.[37] Conversely, in arid regions termite nests mitigate the lack of rainfall and help prevent desertification by increasing ecosystem robustness.[38] One commentary in the journal *Science* explains that one way the mounds do this is by creating "nutrient islands that sustain many other animals besides termites."[39] This is accomplished by concentrating organic matter, nitrogen, and phosphorus in the soil.

Another impressive nest design, constructed by *Macrotermes bellicosus* in Africa, incorporates an arch. Even more wonderful engineering is achieved where they build the two bases of the arch separately and then join them together at the apex, which is possible only if the workers have some sense of three dimensional space.[40] And again, keep in mind that they are doing all this while blind.

J. Scott Turner highlights another impressive aspect of termite mounds: their longevity. A given mound can persist more than ten years, longer than the lifetime of the queen, while workers live for only a couple of months. The implication then is that "a mound is the bequest of one generation of workers to a subsequent generation. The mound is as much a hereditary legacy to the termites occupying it as the strands of DNA inherited from their parents."[41]

Spiderwebs

SPIDERS ARE another of nature's master engineers. About half of known spider species (order Araneae) construct webs made of silk. Spiders can make different types of silk, depending upon its function. For example, the golden orb-weaver spider has seven kinds of silk glands, with six spinnerets.[42] Some is used for spinning webs, of course, but other types are used for wrapping prey and encasing eggs. Silk can be stronger than steel of the same thickness, can stretch more than rubber, and is stickier than most tape.[43] The Goulds describe silk as "easily the most remarkable building material on the planet, and it has one source: arthropods."[44] Despite great effort, humans have yet to produce anything functionally equivalent to silk. Through genetic engineering, attempts have been made to duplicate it without success. The main challenge is replicating the sophisticated and information-rich protein molecules found in the silk produced by spiders and other silk-producing arthropods such as silkworms—proteins that are nearly double the size of average human proteins.[45] Smaller proteins do not have the strength or flexibility of spider silk. Given the advanced genetic and manufacturing technologies available today, it is remarkable that spider silk still cannot be duplicated. This illustrates just how advanced the engineering design of spider silk is.

Orb webs are the most common and familiar types of spiderwebs. A typical garden spiderweb is made of 65 to 195 feet of silk.[46] The webs consist of sticky "catching threads"; radial "spokes" for holding the sticky threads; "bridge threads" that act as guy-lines for holding the web up; "signal threads" that inform the spider through vibrations sensed in the legs that prey is in the web; and "drag lines" for access into the web from her home.[47] The silks employed in the different uses are each unique, being constructed of different combinations of proteins. The types include "slinky" for stretchiness, "zipper" for flexibility, and "lego" for toughness.[48] Construction of the web is a purposeful, goal-driven activity. This becomes particularly obvious as one observes the process in videos available on the internet.[49]

Various spiderwebs, even among spiders of the same species, are far from identical. The most obvious reason for the differences is that each is tailored to its specific location. As the Goulds explain, "Every set of initial anchor points is different; the number of radii is contingent on opportunity; the beginning of the sticky spiral depends on where the longest several radii turn out to be. In short, each web is a custom production."[50] The Goulds postulate that spiders have a form of mapping ability that enables them to implement general design principles in a wide variety of circumstances. This is demonstrated, for instance, by spiders successfully making repairs to damaged webs.

Another source of difference is function. When we think of spiderwebs, we tend to imagine the kind most commonly encountered— the netlike webs spread between trees or attic rafters or walls. But there are various other types, including ones that function as trapdoors into spider burrows, collars that extend out from burrows, and webs that function as tubes on tree bark that can also have hinged doors.[51]

I mentioned signal threads above. They tell a spider that prey is present on the web, but they convey a lot more than just that. Spiders are able to determine both the angle and distance of the prey from the center of the web. They are able to determine the prey location using the same basic technique we use to determine the location of the source of sound. Humans use the difference in intensity of sound received by our ears to estimate the relative location. Spiders do something similar based on the intensity of vibrations received, in their case sensed through eight legs.[52] Obviously the algorithm used in processing information from eight sensors is much more complicated than just the two sensors that humans have. And that's only the half of it. Experiments have demonstrated that spiders can store the coordinates for the locations of at least three different prey trapped in the web.[53]

Providing credible evolutionary explanations for the origin of silk and web design has proven problematic. Several theories have been proposed for the origin of both, but none have been generally accepted.[54] Biologist and spider specialist William Shear concedes that "a functional explanation for the origins of silk and the spinning habit may be impossible to achieve."[55] One complicating factor is that the webs of some spiders that are more distantly related are nearly identical. Shear writes, "It appears probable that several web types are the product of convergent evolution—that is, that the same web has evolved in unrelated species that have adapted to similar environmental circumstances."[56] But as I will argue in Chapter 6, that is an unconvincing explanation for the origin of complex programmed behaviors.

A more fundamental challenge for those seeking to provide a detailed, causally credible explanation for the origin of silk and spiderweb architecture is the number of genes involved in producing silk, and the complex genomes of spiders.[57] After decades of failed attempts to provide a causally adequate explanation, one can be forgiven for concluding that we have no compelling reason to assume that a step-by-step evolutionary pathway to such an information-rich substrate actually exists. And as we will discuss later, there are now some positive reasons to consider that such information-rich systems have for their source something other than a purely blind material process.

Here, suffice it to say that the behaviors and functions associated with both silk and web spinning exhibit many characteristics of human engineering, and engineering of a very high order.

Darwin in the House of Insects

Richard Dawkins, the Oxford biologist and public atheist known for defining biology as the study of things in nature that appear designed but aren't,[58] acknowledges that there is something particularly arresting about animal architecture, even as he sticks to his guns: "Animal artifacts, like caddis and termite houses, birds' nests or mason bee pots, are fascinating," he writes, "but they are a special case among designoid things—an intriguing curiosity."[59]

Similarly, the Goulds limn the marvels of animal architecture but nevertheless insist that these abilities do not require complex programming and instead could have arisen from simple behaviors: "In the end, it is more parsimonious to account for many examples of spider and insect construction by inferring a basic cognitive mapping capacity, combined with an adequate ability to learn and remember, rather than resorting to explanations that require elaborate programming capable of anticipating every conceivable contingency."[60] As the Goulds admit, however, even the more parsimonious of these two hypotheses would require a "huge upgrade in the cognitive status formerly awarded to arthropods,"[61] creatures with brains roughly the size of pinheads.

To be sure, recent research has found evidence that the underlying architectural behaviors in at least some insects with architectural skills are relatively simple. An example is the construction of ant nests. One set of experiments indicates that the coordination of building actions in the ant species *Lasius niger* is achieved primarily through modifying three simple behaviors—digging, shaping a pellet, and dropping the pellet on the soil or existing structure.[62] Control of these behaviors is achieved through the interactions between pheromones and a template based on the ant body size. It is theorized that the resulting complexity of the architectural design is "the result of a self-organised process, which involves the recurring execution of these simple actions and the combination of simple regulatory mechanisms."[63]

Certainly there is here an arresting elegance, but not all is as simple as this description might lead us to believe. The term *self-organization* is just a cover for the part of the explanation without content, one with no identified mechanism or other proposed cause. The story also overlooks the programming of the behavioral responses to the external conditions. While the actual individual behaviors may be simple, they are controlled by a system that results in complex nest designs.

If human engineers and software programmers were called on to replicate this ability in robots controlled by artificial intelligence, they would find themselves forced to extend well beyond basic programming techniques in order to deliver the goods. The same is true of the complex programmed behaviors that appear to underlie sophisticated nest construction among bees, termites, and other architecturally proficient insects.

The evidence for programming is strong, and the sophistication of the programming required to manage these architectural feats, stunning. Here is an explanandum worthy of something more than the vague just-so stories routinely offered up by earnest evolutionists.

6. More Evolutionary Conundrums

> What lies at the heart of every living thing is not a fire, not warm breath, not a "spark of life." It is information, words, instructions... If you want to understand life, don't think about vibrant, throbbing gels and oozes, think about information technology.[1]
>
> — Richard Dawkins

WE HAVE TOUCHED ON SEVERAL CHALLENGES TO THE IDEA THAT complex programmed animal behaviors (CPBs) in animals could have blindly evolved. But there are other challenges not yet touched upon, and challenges touched upon that call for a more careful analysis. Here we shift from chapters that focus on particular categories of complex programmed behavior (i.e., navigation and migration, social behavior, architecture) to a chapter that takes up a series of general challenges to CPB evolution. The first we turn to was briefly considered earlier—convergence.

A Tangled Tree

DARWIN'S THEORY of evolution offers a picture of a gradually branching tree of life, with differences among forms growing over time. But then we get something called convergence. Convergence is roughly equivalent to the cladistics concept of homoplasy, which is defined as "similarity in the characters found in different species that is due to convergent evolution—not common descent."[2] Douglas Futuyma defines convergent evolution as the "evolution of similar features independently in different evolutionary lineages, usually from different antecedent features or by different developmental pathways."[3]

Convergence (and homoplasy) is to be distinguished from homology, which under Darwinian evolution is defined as a common feature or trait in species that share a relatively recent common ancestor. A common example cited for homology is vertebrate forelimbs, which have the same basic design among various species of vertebrates. In this case, the similarity is assumed to be due to common ancestry and, with it, common genetics and common developmental pathways.

On the other hand, animals with "convergent" characteristics are those thought to have evolved the shared characteristics independently, and not from a common ancestor that had the same characteristic or trait. Classic examples of convergent evolution are Australian marsupials and placental mammals in Northern Hemisphere continents. Placental mammals include anteaters, wolves, cats, moles, and mice. The corresponding Australian marsupials include kangaroos, wallabies, and wombats. A PBS educational page neatly summarizes the conventional wisdom among evolutionary biologists on this point:

> Marsupials in Australia and placental mammals in North America provide another example of convergent evolution.... They separated from some common ancestor more than 100 mya, and each lineage continued to evolve independently. Despite this great temporal and geographical separation, marsupials in Australia and placentals in North America have produced varieties of species living in similar habitats with similar ways of life. Their resemblances in overall shape, locomotor mode, and feeding and foraging are superimposed upon different modes of reproduction, the feature that accurately reflects their distinct evolutionary relationships.[4]

Figure 6.1 illustrates an example of a convergent trait. The shared trait, indicated by the triangle, occurs in species B and F. What makes the trait convergent is that it does not occur in a common ancestor of B and F. That means that it appeared independently in these two species without a shared evolutionary pathway.

The appearance of such shared traits presents a conundrum for standard evolutionary theory for multiple reasons. First, when evolutionary scientists construct phylogenetic trees they assume that species which share similar traits do so because they inherited that trait from a common ancestor. But convergent evolution represents instances where that assumption failed, because the trait cannot be explained by common inheritance. This challenges the assumptions which are used to build evolutionary trees. Second, evolutionary biologists often cite the great unlikelihood of the same trait arising in multiple lineages. Richard Dawkins puts it this way: "It is vanishingly improbable that exactly the same evolutionary pathway should ever be travelled twice."[5] Other defenders of evolution believe Dawkins is wrong, and that random mutation and natural selection can produce the same outcome (and explain convergence), as we will see later. However, these other evolutionary explanations will also be shown to be inadequate.

An example of multiple convergences are four species of nectar-feeding birds with a shared trait but without a shared common ancestor possessing that trait. The four species are hummingbirds (Americas), honeyeaters (Australia), honeycreepers (Hawaii), and sunbirds (Africa).[6] All four exhibit the long bills that allow them to feed on the nectar of plants.

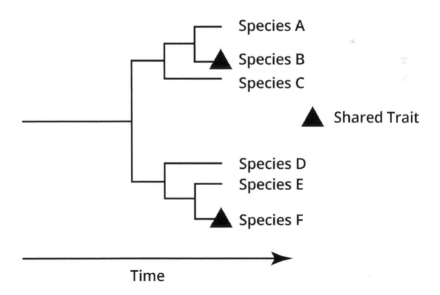

Figure 6.1. Convergent Traits

It is believed that this common trait evolved independently in these four cases—convergent evolution.

As the above example suggests, the idea of convergent evolution isn't restricted to physical traits. It also applies to convergent behavioral traits.[7] In his book *Life's Solution: Inevitable Humans in a Lonely Universe*, Cambridge evolutionary biologist Simon Conway Morris documents numerous apparent convergences, many of which involve shared behaviors.[8] "There is indeed evidence for convergent acquisition of complex behaviour," he writes, and "the role of convergence in the behavioural realm will produce quite a few more surprises."[9]

Why would a blind evolutionary process repeatedly hit upon the same solutions? Rutgers University evolutionary biologist George McGhee provides hundreds of examples of convergence in his aptly titled book *Convergent Evolution*. McGhee explains that convergent evolution was once believed to be produced primarily by functional constraint, where "form follows function." However, McGhee asserts that more recent research has shown that most convergent evolution is due to developmental constraint. "The same forms have been produced by the repeated channeling of evolution along the same developmental trajectory," he writes. "Natural selection has a limited repertoire of potential forms from which to choose, and convergent evolution is the result."[10] According to this thinking, unrelated animals may evolve in highly similar ways and, given enough engineering knowledge about design constraints, we could even predict in many cases what convergent form or convergent behavior might emerge.

But it is one thing to speak vaguely of design constraints. It is another to offer detailed and cogent explanations for various convergences. The latter remain lacking in the scientific literature. Design constraints do not explain all instances of convergence, particularly animal behavior. In many if not most cases, the behavioral traits are contingent and not the result of some constraint.

Here we will focus on convergent behaviors, an area touched on less frequently than convergent physical traits. We will review some of those along with some not previously discussed.

Navigation

MANY OF the navigational abilities described earlier have developed independently in a wide variety of animals. Many animal types have a magnetic sense they use as a compass. It exists in some animals that occupy all major groups of vertebrate animals, mollusks, crustaceans, and insects. Or to take another example, various animals employ the sun as a compass, including butterflies, ants, bees, and birds. This is quite a trick, involving as it does an algorithm to calculate the compensation for the movement of the sun across the sky.

Some animals are also able to use the polarized light from the sun as a compass source. Animals known to be able to detect polarized light include honey bees, ants, crickets, and locusts. Another diverse group of animals have a map sense that enables true navigation. This includes birds, sea turtles, lobsters, and mollusks.

If all or most animal species had these navigational tools, one might simply attribute it to common descent and explain the occasional outlier without the ability as the victim of "devolution"—that is, the animal lost the ability somewhere in its evolutionary history. But instead we have some species possessing these abilities while many more in the evolutionary neighborhood, including the (supposed) common ancestors, apparently lack them, thus forcing evolutionists to appeal to convergent evolution again and again.

Architecture

VARIOUS INSTANCES of nest design across the animal kingdom also bespeak convergence. A particularly striking example is between weaver ants and birds that weave or sew nests. Both use complex weaving methods to engineer their nests. Another convergence is that nest weaving occurs in four different distantly related groups of ants. Construction of temperature-regulated nests is yet another convergent behavior which occurs in bee hives, termite mounds, and some bird nests.

Social Behavior

SOCIAL BEHAVIOR is thought to have originated independently in Hymenoptera (which includes ants, bees, and wasps), as well as in their distantly related cousins, termites, which belong to a different clade. (A clade is a group that consists of a common ancestor along with its various descendant branches.) E. O. Wilson compared the social behavior of the social Hymenopterans to termites,[11] noting several differences but also many similarities. Among the similarities are the caste structures, the use of chemical trails, caste inhibition through pheromones, shared grooming between individuals, nest structure complexity, and sharing of food and liquids (trophallaxis). There is also commonality in the algorithmic mechanisms that control expression of the social behaviors. Another common behavior in ants and termites is farming, an extremely unusual phenomenon in the animal kingdom. Yet these behaviors are not thought to derive from a common ancestor.

Evolution and Convergence

ALL OF the above behaviors greatly enhance an animal's ability to survive and thrive. However, a trait's providing an important function does not explain how it evolved independently numerous times throughout the animal kingdom. As the old adage goes, explaining "survival of the fittest" is different from accounting for the "arrival of the fittest." And as one biologist noted in the journal *Nature*, "The modern synthesis is remarkably good at modelling the survival of the fittest, but not good at modelling the arrival of the fittest."[12]

Darwin asserted that natural selection explains similar organs that were not present in a common ancestor. "As two men have sometimes independently hit on the same invention," he wrote, "so natural selection, working for the good of each being and taking advantage of analogous variations, has sometimes modified in very nearly the same manner two parts in two organic beings, which owe but little of their structure in common to inheritance from the same ancestor."[13] The great twentieth-century Harvard evolutionary biologist Ernst Mayr agreed with Darwin that the power of natural selection explains convergent evolution: "Convergence illustrates beautifully how selection is able to make use of the intrinsic variability of organisms to engineer adapted types for almost any kind of environmental niche."[14] But again, do such claims really explain *how* these complex traits *arise* in the first place, and over and over again? If we are to follow the evidence, we must not assume but rather ask whether the repeated evolution of the same complex trait is likely to occur given the mechanism of random mutation and natural selection.

A major theme of Morris's book *Life's Solution* is that the widespread appearance of biological convergence is evidence of the constraints on evolution. "Convergence is ubiquitous and the constraints of life make the emergence of the various biological properties very probable, if not inevitable," he writes.[15] Morris strongly disagrees with the view of Stephen Jay Gould that the history of evolution is filled with contingent events, explaining that his book is meant to "refute the notion of the 'dominance of contingency.'"[16] Regarding social behavior, he comments, "There is indeed evidence for convergent acquisition of complex behaviour," and he concludes that "the repeated rise of such societies, and at least evidence of the displacement and ultimate extinction of less successful equivalents, suggests that such an arrangement is a biological inevitability."[17]

However, the evidence indicates it is not inevitable, but rather contingent, because only a few groups of the numerous species of ants, bees, and wasps are social. And indeed, unlike some physical traits, such as wing designs or fish body and fin design, there is no fundamental principle that would make many CPBs deterministic. Indeed, and this bears repeating, if one is not wedded to a materialistic version of evolution, there is no reason why blind evolution would render their repeated, independent appearance probable or expected.

Various flying animals share common features in wing design. The technical reason is that efficient wings have specific characteristics that maximize lift and minimize drag, which imposes a constraint on the design of wings that have sufficient performance for flight. Analogous reasons apply to the commonality in the design of fish fins. But with most behaviors there are almost always multiple dramatically distinct solutions available, for behavioral solutions do not have the kind of constraints that physical characteristics do. Any number of behaviors might meet a particular need or function. Examples include the various types of navigation sensors and functions in animals. There are also numerous migration strategies.

The same is true for social behaviors, architecture, and communication systems across various insect species. While they share much in common in terms of colony functions and benefits, there is significant variation in how the functions are implemented. That is seen in the differences between bee, ant, and termite social colonies. What this means, again, is that the appearances of specific complex programmed behaviors are not highly constrained and inevitable, but contingent. Thus, one cannot appeal to design constraints plus natural selection even to begin to explain the widely various CPBs we find in the animal kingdom. Instead, we must look elsewhere for a satisfactory explanation.

Multiple evolutionary biologists have invoked concepts drawn from evolutionary developmental biology ("evo-devo") to explain convergence in animal physiology.[18] Entymologists Amy Toth and Gene Robinson extend evo-devo to explain the convergent evolution of social behavior.[19] As evidence for this model, they cite the possible presence of common genetic toolkits in various social insects. This hypothesis is very similar to the groundplan hypothesis for the origin of social behavior described in Chapter 4. European biologist Rinaldo Bertossa also supports the evo-devo view that common genetic and neuronal structures foster the development of similar behaviors. He asserts that there is evidence of the conservation of neuronal circuits across different species, and that "functional differences could arise by respecification of common circuits."[20]

It is possible that common neuronal circuits can explain the convergence of relatively simple behaviors across species, as proposed by Bertossa. However, as discussed in Chapter 4, evidence suggests there are a large number of genes and additional epigenetic mechanisms involved in social behavior. And these genes and other mechanisms are often unique to the species that are social. This observation conflicts with the notion of a common toolkit consisting of a small number of genes.

Another problem with the evo-devo explanation is that many convergent behaviors, such as navigation and communication, involve systems engineering. Evo-devo, a blind process, does not produce coordinated, top-down design, and thus is a poor candidate for addressing the many aspects involved in the systems engineering of behavior.

Even the staunch evolutionary biologist Jerry Coyne, of the University of Chicago, admits the limitations of appeals to convergent evolution. Yes, there are many convergences between marsupials and placental mammals, but as Coyne notes, "There are also many types of mammals that evolved elsewhere that have no equivalents among marsupials. There is no marsupial counterpart to a bat (that is, a flying pouched mammal), or to giraffes and elephants. Most tellingly, Australia evolved no counterpart to primates, or any creature with primatelike intelligence."[21] Such counterexamples are evidence that the appearance of species that do exhibit common characteristics is largely contingent rather than deterministic.

Design and Teleology of Convergent Behavior

EVERY CATEGORY of complex programmed behavior discussed in this book includes common behaviors spread among animal species which do not share a common ancestor that exhibited the behavior. These behaviors appear to be convergent, as that term is usually applied. Convergence is just a word used to describe such apparent cases. It does not explain the origin of the phenomenon. The most prominent evolutionary explanations for convergence are design constraints and selection pressure. For example, McGhee asserts that "the a priori standards of nature are mindless functional constraints imposed by the laws of physics and geometry."[22] That is, for certain functions there are said to be a very limited range of engineering designs that work at all (design constraints), and among those, some are so superior to the others that selection pressure tends to push the evolutionary process toward the superior options to the exclusion of the inferior but workable ones. Obviously some would-be solutions work while others do not—but this truism in no way addresses the complexity of the workable solutions and whether they are likely to evolve via unguided processes such as random mutation and natural selection. Vaguely invoking factors such as functional or physical constraints in no way addresses the specific mutational pathways by which a complex biological system might arise. Invoking such factors as an explanation is, again, to conflate survival with arrival.

French behavioral ecologist Luc-Alain Giraldeau writes, "That natural selection causes evolution toward enhanced design of organs and behavior can be inferred through examples of convergent evolution."[23] But he appears to support the Darwinian evolution of biological features not because this seems likely based on what we know about biological complexity, but simply because these complex features repeatedly appear throughout nature and he's a committed evolutionist. Blind evolution is his hammer, so convergence must be a nail.

To be clear, the argument typically offered is not as simplistic as "evolution causes convergences, because evolutionary theory is true." The actual argument goes something like this:

1. Certain forms of behavior are highly adaptive.
2. Because these forms of behavior are highly adaptive, they are favored by natural selection.
3. This creates selection pressure for these behaviors.
4. Therefore, such behaviors evolve independently in different groups of animals.

By way of summary, there are here two problems. First, the many counterexamples to behavioral convergence suggest that contingency, rather than necessity, is the order of the day. Second, and most fundamentally, natural selection can only select for traits that appear in the first place. Therefore, random variation of one form or other would have to first cause the appropriate genetic changes to occur that produce the behavior. This is rarely addressed by proponents of convergent evolution, but absolutely needs to be, since as we have seen, randomness, even coupled with natural selection, is a poor candidate for creating in an orchestrated fashion the numerous coordinated changes needed to provide any sort of fitness advantage.

Evolving and Crashing CPBs

LET'S DELVE further into this question of fitness advantage. A central challenge facing the idea that complex programmed behaviors evolved through a gradual and blind process of trial-and-error evolution concerns the functional nature of CPBs. The gradual evolution of a novel complex programmed behavior would require changes in multiple elements of the phenotype involved in the control of CPBs—for example, 1) physiology, 2) brain structure, 3) sensors, and 4) the brain algorithm.[24] Since at least those four elements are involved in controlling CPBs, genetic variations and/or mutations that cause a change in the programmed behavior have to at least be compatible with those four elements and allow functional coherence. A genetic change in one element that is incompatible with another might result in no behavioral change, or in a maladaptive behavior. In the worst cases it would, in essence, crash the CPB.

That at least four phenotypic elements are involved also means multiple changes typically must occur to enable the new behavior to arise. While the changes do not necessarily have to occur simultaneously, they at least need to occur in a coordinated manner that is coherent and adaptive. This is required to be consistent with Darwin's thesis that each incremental change must be beneficial.

The problem is, the likelihood of multiple coordinated genetic changes is extremely low, as described in Chapters 3 and 4 concerning migratory and navigation behaviors, and insect social behavior. Navigation and migratory behaviors involve many complex traits that are highly integrated. Conceptually it is implausible for all of these traits (navigation physiology, navigation algorithms, migratory physiology, migration algorithms) to develop in a coordinated manner through random mutations and natural selection. In the case of social communications, for pheromone signals to be effective the receiver must be able to detect and discriminate among pheromones. This is no small feat when the animal employs as many different pheromone compounds as in, for example, the carpenter ant. The sensing mechanism must be highly sophisticated to discriminate all combinations of compounds. Tristram Wyatt writes, "Generally speaking, pheromones do not require learning: they seem to be 'innate,' 'hardwired,' predisposed, or 'work out of the box.'"[25]

A key problem in explaining the evolution of pheromone communications is the synchronization of sender and receiver. The information from the sender must be received and decoded by the receiver to have the same meaning. In insects, the genetics of the sending and receiving of pheromones are rarely linked, often residing on different chromosomes.[26] Since they are independent, coordination of the changes is unlikely.

Neutral Evolution to the Rescue?

SOME HAVE proposed that neutral mutations accumulate and eventually result in novel functional traits such as CPBs. The neutral theory of molecular evolution was first proposed by Japanese biologist Motoo Kimura. A central claim of the theory is that neutral mutations—changes which provide neither an advantage nor a disadvantage to the organisms—can accumulate without being subject to natural selection, resulting in random genetic "drift" over time. Michael Lynch, a supporter of the theory, writes, "Numerous aspects of genomic architecture, gene structure, and developmental pathways are difficult to explain without invoking the nonadaptive forces of genetic drift and mutation."[27]

However, the theory remains controversial. Biologists Andrew Kern and Matthew Hahn comment that genomic research since the neutral theory was introduced fifty years ago has demonstrated that "each of the original lines of evidence for the neutral theory are now falsified."[28]

They conclude that neutral mutations are not a major factor in evolution because natural selection has had the predominant role in genetic variation. Lynch and a group of colleagues responded in the journal *Evolution*, asserting that the evidence shows that

> (1) A large fraction of the genome of organisms studied to date is subject to mutations that are effectively neutral with respect to their fitness effects, and hence evolve under genetic drift. (2) The great majority of newly arising mutations that do affect fitness (i.e., non-neutral mutations) are deleterious, and the predominant mode of natural selection is purifying in nature, removing these deleterious mutations from populations.[29]

What the two sides seem to ignore in this tug-of-war is the possibility that neither neutral evolution nor evolution via natural selection can generate fundamentally novel structures and novel behaviors that are functional. Neutral mutations do occur and can accumulate in genomes. However, those that are truly neutral are not positively selected, which means there is no reason for natural selection to preserve specific changes that are needed for some trait. Coyne thus observes that genetic drift is "powerless to create adaptations" and "the influence of this process on important evolutionary change… is probably minor, because it does not have the molding power of natural selection. Natural selection remains the only process that can produce adaptation."[30] The implication is that neutral mutations do not play any significant role in the origin of CPBs.

So where are we left, as regards the chances of either neutral evolution or evolution by natural selection? We are left in a quandary. Only a small minority of mutations that occur are beneficial and can be selected. But the development of a CPB requires a large number of individual beneficial mutations that survive selection. And as we saw above, the challenge is compounded by the fact that beneficial mutations must, in many cases, be coordinated with other beneficial mutations to be immediately beneficial.

Which Came First?

It's easy to overlook the coordination problem, but if one is to develop a realistic evolutionary model for the origin of CPBs, the problem must be faced head on. If we imagine that the phenotypic changes involved in CPBs are not simultaneous, the question arises: Which changed first: morphology (form) or function (behavior)? Darwin wrestled with this issue, writing:

> It has been objected to the foregoing view of the origin of instincts that "the variations of structure and of instinct must have been simultaneous and accurately adjusted to each other, as a modification in the one without an immediate corresponding change in the other would have been fatal"... The force of this objection rests entirely on the assumption that the changes in the instincts and structure are abrupt.[31]

In other words, Darwin answers the challenge by asserting that the changes to both behavior and structure were gradual. He admitted that his explanation left much still unexplained, but dismissed this lacuna as inconsequential: "It is difficult to tell, and immaterial for us, whether habits generally change first and structure afterwards; or whether slight modifications of structure lead to changed habits; both probably often change almost simultaneously."[32]

However, the conundrum is far from inconsequential, for both of Darwin's proposed scenarios here are inadequate. If one precedes the other, then we are simply back to the objection Darwin raised but failed to answer. If structure and instinct change nearly simultaneously—that is, together in a very gradual back-and-forth process on the way to the fundamentally new structure/instinct, then the devil is in the details. How specifically can evolution by natural selection gradually traverse not just one daunting transitional gulf but two simultaneously? What detailed evolutionary pathways are being offered, either by Darwin or by any of the current advocates of neo-Darwinism, concerning any complex programmed behavior in the animal kingdom? Such explanations are virtually never forthcoming, probably because, as explained in Chapter

3, the probability of evolution producing numerous multiple coordinated genetic changes is extremely remote.

As we have seen, many behaviors depend upon appropriate physiologies; otherwise they cannot be performed. Conversely, a unique physiology can be detrimental to adaptation (energetically expensive for example) if an associated behavior is not present. For both the behavior and physiological trait to be adaptive, both must be instantiated at the same time. In the case of navigation and migration, it would be maladaptive for an animal to have a complex, energetically expensive navigation system if it never migrated, meaning that evolution would tend to select against it or, where genetic drift predominated, pay it no favors where it began to evolve by dumb luck.

A counterargument is that some bird and insect species never migrate, yet all have some capabilities to do so. Doesn't this show organisms retaining traits that don't give an advantage? It seems true that apparently all birds have basic navigation capabilities. However, populations change their migratory propensity depending upon environmental conditions. Therefore, it is still adaptive to maintain the capability over the long haul, and says nothing about the origin of these navigation systems. A similar situation concerns the ability to fly. Possessing a suite of adaptions necessary for flight is extremely expensive in terms of tradeoffs, compensated for only by the actual ability to fly. Yet there are some bird and insect species that are flightless while maintaining wings. These, however, are examples of devolution, where a once functional capability is lost.[33] Such examples offer no help in explaining the origin of novel, functional traits.

Evolving Master Algorithms

FUNCTIONAL INFORMATION is the common element necessary for creating the complex programmed behaviors explored in these pages. Such information is essential, both in defining the underlying algorithms and for processing and applying external information (sensory data) to the various functions.

Navigation, for instance, involves several functions that require information, such as programming for algorithms that process navigation sensor data (e.g., algorithms for magnetic compass, sun compass, odometers, and path integration). In cases where the location of the migration destination is programmed, information defining direction and distance is needed. In some cases there even seems to be information defining a map.

In architecture, nest building requires information that defines the architectural blueprint. Various behaviors of social insects, such as the division of labor, require information to define the contexts and signals where each behavior is performed. All elements involved in social communication require information, such as the proteins that define pheromones and the coding and protocols of signals such as alarm calls and the honey bee dance. Programmed learning requires algorithms that control the process, including defining what information is used and how it causes changes in behavior. In addition, the algorithms that control these behaviors must be encoded in neural mechanisms, as illustrated in Figure 6.2.

Implementing the algorithms that control these behaviors involves several different levels. In describing a closely related aspect of animal behavior, neuroethologist Geoffrey Adams and his co-authors describe three levels where decision-making is implemented and can be analyzed: computational, algorithmic, and implementational.[34] In the authors' terminology the computational level can be equated with taking the source of the information needed to solve the behavioral strategy and inputting it into the algorithm. The algorithmic level defines the strategies for how the desired behavior is solved. The implementational level concerns how the algorithms are instantiated, primarily through neural mechanisms.

The authors further note that implementing the algorithms can occur through various means and may involve "neuronal or genetic circuits, neuromodulators, hormones, or a complex interplay of these."[35] They explain that a genetic circuit and a neural network may implement the same algorithm, though the biological details remain distinct. In other

Figure 6.2. Implementing a Behavior Algorithm

words, such algorithms may be functionally equivalent but implemented through quite different mechanisms.

In fact, we still know very little about the exact mechanisms by which such algorithms are implemented. While many believe this is accomplished primarily through neural networks, these are poorly understood, except in the most basic cases.

It was believed that basic behaviors such as associative learning involved relatively simple mechanisms. An excellent example of a "simple" behavior involves habituation and sensitization in invertebrates. Habituation has been studied extensively in the marine snail *Aplysia californica* by Nobel Prize recipient Eric Kandel, because of what was thought to involve a relatively simple neural mechanism that controls its gill-withdrawal-reflex (GWR).[36] Early research and modeling indicated that habituation occurs as a result of the change in response at the synapses of motor neurons. However, more recent research revealed that the mechanism responsible for the habituation and dishabituation is actually not so simple. The response in *Aplysia* is accomplished through a network of approximately three hundred neurons, including sensory neurons, motor neurons, and interneurons.[37]

Previous models predicted that the gill-withdrawal-reflex involved a simple stereotyped motor behavior mediated by a relatively simple identified neural circuit. However, biologists Janet Leonard and John Edstrom note that "it is now clear that the operationally defined GWR, the response of the gill to weak tactile stimulation of the siphon, is not a simple stereotyped reflex but rather involves a heterogeneous set of action patterns."[38] They add that the results of thirty years of research on the gill-withdrawal-reflex and its neural control have led them to conclude "that this simple system, like many others, is much more complex and

variable than had been initially predicted."[39] Given that such simple behaviors involve relatively complex neural mechanisms, it is obvious that
much more complex programmed behaviors correspondingly involve significantly more complex neural and related mechanisms.

The evolution of algorithms useful for solving ubiquitous problems
facing animals that exhibit complex programmed behaviors are, according to Adams and his colleagues, "likely to be products of convergent
evolution, primarily because they are robust, require only simple components, and do not require a centralized architecture. Such algorithms are
thus more likely to be repeated across taxa."[40] But as we have seen, these
behaviors are complex rather than simple. Additionally, each is unique,
with much that is distinct from other complex programmed behaviors
in the same grouping (e.g., navigation/migration). Therefore, the algorithms that control these behaviors are not simple and convergent but
instead sophisticated and unique, with specialized solutions. Thus, convergent evolution is both an inadequate explanation for such sophisticated algorithms and not a very relevant explanation, since there is much
less convergence among the individual instances of these complex behaviors than it would seem at first blush.

There are three fundamental ways an animal can obtain information
used in determining behavior: 1) The information is inherited, either in
the genome or through epigenetics; 2) the information is gotten from
experience; or 3) the information is gotten through communication with
other members of the species.[41] All three play roles in the full range of
animal behavior. As we have seen, it is not simply a question of how information is embedded within the genome or transmitted epigenetically.
It is also a question of the origin of information that programs learning. The question concerning all three ways of obtaining information is,
*What is the ultimate source of the information that defines the algorithms
and neural mechanisms that control these behaviors?* That is still a mystery.

Darwinian Evolution and Biological Information

ACCORDING TO Brown University biologist Kenneth Miller, an outspoken defender of evolution, biological information is generated by "just three things: selection, replication, and mutation.... Where the information 'comes from' is, in fact, from the selective process itself."[42] Similarly, Stony Brook University evolutionary biologist Douglas Futuyma writes, "The apparent goal-directedness is caused by the operation of a program—coded or prearranged information, residing in DNA sequences—that controls a process." He further explains that the three basic constituents of evolution by natural selection needed to produce these features are variation, differential fitness, and heritability.[43] Richard Dawkins draws out the wider implications of this view:

> All appearances to the contrary, the only watchmaker in nature is the blind forces of physics, albeit deployed in a very special way. A true watchmaker has foresight: he designs his cogs and springs, and plans their interconnections, with a future purpose in his mind's eye. Natural selection, the blind, unconscious, automatic process which Darwin discovered, and which we now know is the explanation for the existence and apparently purposeless form of all life, has no purpose in mind. It has no mind and no mind's eye. It does not plan for the future. It has no vision, no foresight, no sight at all. If it can be said to play the role of watchmaker in nature, it is the *blind* watchmaker.[44]

Some (Darwin included) view natural selection as a natural law, which, if the case, would allow Darwin's theory to meet what many view as a criterion of any good scientific theory, namely that it be grounded in natural law.[45] Michael Ruse explains how this fits into the broader rule known as methodological naturalism: "The methodological naturalist is the person who assumes that the world runs according to unbroken law; that humans can understand the world in terms of this law; and that science involves just such understanding without any reference to extra or supernatural forces like God."[46] Examples of natural laws include the law of gravitation, Boyle's Law (describing the relation between the pressure and volume of a gas), Einstein's equation relating mass and energy ($E = mc^2$), and Newton's laws of motion. Another example is the first law of thermodynamics, which concerns the conservation of energy and states that "energy cannot be created or destroyed in an isolated system."

The claim of many Darwinians is that biological phenomena, and more specifically natural selection, can be classified in the same category as these sorts of laws. However, the examples cited above are based on observations of ongoing physical regularities. In contrast, and as philosopher of biology Elliott Sober explains, "Many of the general laws in evolutionary biology seem to be a nonempirical mathematical truth."[47] He cites the example of the Hardy-Weinberg principle of gene allele frequency, which is sometimes referred to as a law. As Sober points out, it is simply based on the mathematical distribution of alleles in a population, given certain assumptions. Similarly, concerning Mendel's laws of inheritance, Mayr writes, "The so-called laws of biology are not the universal laws of classical physics but are simply high-level generalizations."[48] There are no physical deterministic processes that underlie these observations. Therefore, despite the claim that evolution can be explained by natural laws, the truth is that there are no actual natural laws in biology that are equivalent to those in physics. Sober concedes that "if we use the term 'tautology' sufficiently loosely (so that it encompasses mathematical truths), then many of the generalizations in evolutionary theory are tautologies. We seem to have found a difference between physics and biology. Physical laws are often empirical, but general models in evolutionary theory typically are not."[49]

But set aside now the question of whether the Darwinian mechanism qualifies as a natural law. Let's return to the more fundamental question: Can it or any natural law generate the biological information necessary for the diversity of life? According to origin-of-life researcher Manfred Eigen, information such as we find in DNA can come from natural laws. Eigen agrees that the key to the origin of biological complexity is information. The challenge, he writes, is to "find an algorithm, a natural law that leads to the origin of information."[50] He describes the typical evolutionary explanation for the origin of biological information:

So how has the information in genetic blueprints, the fixation of particular arrangements of symbols, come into being? A biologist would answer: by natural selection! He would add that the gene sequences contained in organisms, coding as they do for functions that are optimally adapted for life, are in fact the products of a whole series of changes in the sequence, stabilized one after another by selection. Darwin's principle brings about what theoreticians would call the generation of information.[51]

Eigen continues: "It will have to be a dynamic principle. Information arises from non-information… Evolution as a whole is the steady generation of information—information that is written down in the genes of living organisms."[52]

That is a very strong assertion. Can supporters of this assertion provide evidence to support it? Mathematician and philosopher William Dembski argues that regular natural laws cannot generate the sort of information we find in DNA.[53] His more technical terms for such information is *functional information* or *complex specified information*. This is the sort of information the average person thinks of when one hears the term information—the sort of stuff we find in books, software programs, and DNA. Laws of nature involve repeating patterns, but information such as we find in books, software programs, and DNA is aperiodic and largely non-compressible. Consider the following string of letters:

abcabcabcabcabcabcabcabcabcabcabcabcabcabcabcabcabcabc

That string of letters is fifty-four characters long but is easily communicated in compressed form, namely "Repeat abc eighteen times." It's just a simple repetition. It's not complex in the way the rest of the letters/words in this paragraph are. Law-like patterns can produce things like "abc repeating," but according to Dembski, they cannot produce novel information of the sort found in the meaningful sentences in this book or the functional code in a software program. As Stephen Meyer explains, "Laws describe highly predictable and regular conjunctions of events—repetitive patterns, redundant order. They do not describe the kind of complexity necessary to convey information."[54]

We know this from our uniform experience and from reason. Whatever cause is responsible for generating complex specified information such as we find in software programs and DNA needs the freedom to choose based on a functional goal. The cause must be able to generate contingent rather than deterministic outcomes. "Laws can shift information around or lose it, as when data gets compressed," Dembski writes. "What laws cannot do is produce contingency; and without contingency they cannot generate information."[55]

What about random processes, such as the chance mutations said to work alongside natural selection in the neo-Darwinian process? Dembski shows through an application of probability mathematics that random processes cannot generate enough specified information to make a blind evolutionary process viable, even when yoked to natural selection. A million monkeys banging away on a million typewriters for a million billion years will never stumble upon a Shakespearian quatrain, much less a Shakespearian sonnet or play, and chance mutations will never create fundamentally new and functional protein folds, much less whole new organs and body plans. The odds are just too long. Dembski's argument has been corroborated by laboratory work on protein folds by Douglas Axe and others.[56]

Dembski laid the groundwork for his analysis in his Cambridge University Press monograph *The Design Inference*, and then addressed additional problems associated with Darwinian explanations for the origin of information in later books and papers.[57] As the title of his book *No Free Lunch* suggests, you can't get something for nothing. Dembski identifies "the central problem of biology" as "the origin of complex specified information…. Where, then, does complex specified information or CSI come from, and where is it capable of coming from?"[58]

Several computer simulations have been developed that purport to model blind evolution and claim to demonstrate an increase of information. Kenneth Miller cites one such simulation as a good example of how neo-Darwinian evolution creates information.[59] Again, according to Miller, the only things needed to accomplish this are replication, mutation, and selection.[60] The purpose of these programs is to demonstrate that Darwinian evolutionary processes could indeed have generated the biological information essential for the vast diversity of life on earth.

The question is whether these models are accurate demonstrations of blind evolution and real biological processes. With any computer model that purports to represent a real-world phenomenon, there is always a question of fidelity. A mantra with engineers is, "In theory, theory and reality are the same. In reality they are not."[61] Much depends on the complexity of the natural phenomenon that is being modeled. Predicting the weather is done with sophisticated computer models. While short-range forecasting is fairly accurate, longer range predictions are much less accurate. An example is the complexity is predicting hurricane tracks.

Recent research in a field called "evolutionary informatics" analyzes the design and performance of evolutionary computer models. According to Dembski and his colleagues, computer scientists Robert J. Marks II and Winston Ewert, the analysis has found that the fundamental design of evolutionary models is the same as those used by engineers and computer scientists, where "there is always a teleological goal imposed by an omnipotent programmer, a fitness associated with the goal, a source of active information, and stochastic updates."[62]

In other words, the programs that purport to model a natural evolutionary process that is blind and purposeless, without target, are themselves thoroughly purposive and goal-directed. The algorithms are the product of creative intelligence, being programmed to select a predetermined outcome. In contrast, true natural selection is not based on a designed algorithm, and as many advocates of neo-Darwinian evolution have stated, it has no predetermined outcome or target.

Also, the evolutionary models only *appear* to generate new information, when in fact the information already existed in the program in the form of specified algorithms. Such algorithms search for specific outcomes or targets. However, to do so each model "requires sources of knowledge to generate active information to guide the search."[63]

Another problem with such models is that they include assumptions that are not consistent with biology. One such model assumes that after each mutation the result is compared to a target and its overall fitness evaluated, with more matches meaning higher fitness and therefore preservation for the next round. Dembski and his colleagues comment that the result is "equivalent to assuming not just that all loci can be the sites of beneficial change independently, but also that these independent benefits are additive. Both of these assumptions fail to square with biology."[64] Such unrealistic assumptions have the effect of drastically reducing the apparent time for evolution to result in complex adaptive traits. In light of this, Dembski, Marks, and Ewert concluded that "undirected Darwinian evolution has neither the time nor computational resources to design anything of even moderate complexity. External knowledge is needed."[65]

This analysis of evolutionary computer models comports with findings from observation of changes in various microbes over time, as well as from knockout experiments on fruit flies and other species. As biologists Tom Strachan and Andrew Read note, "Making random changes in a gene is quite likely to stop it from working, but very unlikely to give it a novel function."[66]

This has been corroborated by the now-famous long-term evolution experiment conducted by Michigan State University microbiologist Richard Lenski on *E. coli* bacteria. The experiment has run for over 70,000 generations of *E. coli*, and the results have shown the difficulty and improbability of random mutations plus natural selection generating novel biological forms. There have been a small handful of interesting developments among the *E. coli*, but when investigated, it was found that any helpful adaptation was managed by a mutation breaking something in the *E. coli* bacteria rather than by building anything new—akin to making a door more burglar-proof by breaking the lock so that it's permanently stuck in the locked position.[67]

After reviewing this experiment as well as studies of evolution in other microbes and viruses, such as HIV and the malaria parasite, biologist Michael Behe concluded, "Modern research reveals that… not only are random mutation and natural selection grossly inadequate to build complex structures, they strongly tend to break them."[68]

The Origin of the Information Defining CPBs

THE ABOVE analysis suggests that mindless evolutionary processes are powerless to create the genetic information essential to complex programmed behaviors. But that's only part of the problem facing evolutionary theory. As noted, there is evidence that the information needed to program complex programmed behaviors cannot simply reside in the DNA that codes for proteins.

CPBs, recall, involve not just the behaviors but complex physical structures, and not just "software" but also "wetware." So, for example, the "software" that controls complex programmed behaviors must reside in algorithms that become programmed in the brain. It might be that this programming resides in the non-coding portion of the genome, or is somehow controlled through epigenetics.

Non-coding DNA is what used to be referred to as "junk DNA." Advocates of neo-Darwinian evolution commonly held that elements of genomes of organisms that do not code for proteins were without function—the flotsam and jetsam of evolution's mindless trial-and-error process. But research conducted under the Encyclopedia of DNA Elements Project (ENCODE) revealed that much of the non-coding DNA is functional after all. In the case of humans, "nearly 99 percent of the ~ 3.3 billion nucleotides that constitute the human genome do not code for proteins," and thus were presumably non-functional.[69] However, the initial estimate from the ENCODE project is that as much as 80 percent is biochemically functional.[70] And evidence has continued to accumulate to support the idea that "noncoding regions of the human genome harbor a rich array of functionally significant elements with diverse gene regulatory and other functions."[71] That opens the door to the possibility that some of the functionality of non-coding DNA is related to an animal's behavior. A study discussed in Chapter 4 found evidence of a role for non-coding DNA in the development of social behavior in bees.[72]

What is the essence of all this biological information? Philosopher Robert Stalnaker of MIT describes the basics of information: "To learn something, to acquire information, is to rule out possibilities. To understand the information conveyed in a communication is to know what possibilities would be excluded by its truth."[73] Understood in this way, biological information defines the specific algorithms that control CPBs while excluding alternative algorithms. CPBs such as navigation algorithms are both complex and specific and defined by information that excludes numerous alternatives.

One can compare the algorithms underlying artificial intelligence (AI) with the algorithms that control complex programmed behaviors in animals. Conceptually they have similar functions and capabilities in perception, communication, information processing, learning, and decision making. Recall that the mathematical definition of an algorithm is "a formal procedure for any mathematical operation, especially a set of well-defined rules for solving a problem in a finite number of steps."[74] The concept came first, then computers were developed that could run algorithms written by human programmers.

Today we find algorithms being used all around us. Examples of algorithms are found in internet search engines such as Google, which use highly sophisticated algorithms that search the internet for any term that users query. Smartphones include algorithms in GPS route navigation, voice recognition, and various other applications. Route navigation applications employ complex algorithms to compute the most efficient route to the desired destination. In most cases, there are several possible routes, and the algorithm determines which one is likely to be the fastest. Such algorithms are, of course, never the result of a blind material process.

Another significant problem with Darwinian explanations for the algorithms underlying complex programmed animal behavior is that such programs must follow highly structured processes to function properly, and require the integration of functions in order to work coherently. Similarly, software development processes that are used to code complex algorithms must also be highly structured and coherent. When they aren't, the result is usually software that contains a lot of bugs that compromise software execution. Anyone who has ever written a software program is familiar with this issue.

In addition, random changes to complex algorithms are much more likely to be detrimental rather than beneficial. Think of making such random changes to a computer program, for example an AI software program. It is far more likely to result in detrimental or even loss of functionality than in some improvement or even new function.[75] Similarly, the process of random mutations nearly always results in no change to functionality (neutral) or degradation of the genome and functionality. In rare cases there are mutations that break something in a way that leads to some niche advantage for a population.[76] What we don't see are random changes building fundamentally new forms and complex functional behaviors.

These limitations on evolution are especially true for the algorithms that control CPBs. Because of the complexity of these algorithms, random mutations will degrade the functionality of the algorithm the vast majority of the time, just as do random changes to sophisticated software programs. Furthermore, this argument applies to all evolutionary mechanisms that have been proposed that involve mutation and natural selection. It does not matter whether the changes occur in genes, noncoding DNA, gene regulatory networks, or other epigenetic mechanisms. The same problem is inherent in any such mechanism where the algorithms for CPBs might be defined.

Complex programmed behaviors are contingent, meaning they are not necessary or deterministic. Thus, the information that defines these behaviors is also contingent. Dembski proposes a principle called the conservation of information, which holds that a process cannot produce any more information than was put into it.[77] This is equivalent to the conservation of energy defined by the first law of thermodynamics. If we take a computer program as an example, any program cannot output more information than was put into the program initially. Some insist that certain computer algorithms can generate more information than was put into them. But as Marks, Dembski, and Ewert show in painstaking detail in *Evolutionary Informatics*, this isn't the case. Just as you can't get "something from nothing," the information that is output by an algorithm always reflects the information programmed into the algorithm.[78]

If such processes are incapable of creating information, then what is? We turn to that question in the next chapter.

7. Complex Programmed Behaviors—Intelligently Designed

> The fundamental claim of intelligent design is there are natural systems that cannot be adequately explained in terms of undirected natural forces and that exhibit features which in any other circumstance we would attribute to intelligence.[1]
>
> — William Dembski

We often hear talk of "the scientific method," as if there is a single unitary methodology across all branches of science. In fact philosophers of science generally insist that there is no single methodology that is accepted as superior for all branches of science.[2] The method most appropriate for assessing historical scientific hypotheses such as neo-Darwinian evolution is called "inference to the best explanation," as described in the book of the same title by Peter Lipton.[3]

There are three methods of making an inference—deduction, induction, and abduction. Deduction, or logical deduction, applies the rules of formal logic. Mathematical reasoning is almost always done through logical deduction. A simple example of mathematical deductive logic is:

1. $A = B$
2. $B = C$
3. Therefore: $A = C$

With deduction, the conclusion is necessarily true, if the premises are true. In the example, if both 1) and 2) are true, then 3) is necessarily true. Another example is:

1. All humans have 23 pairs of chromosomes

2. Mary is a human being

3. Therefore: Mary has 23 pairs of chromosomes

Inductive inferences typically are based on statistical analysis, which increases the likelihood that the conclusion is true, but does not guarantee the truth of the conclusion. An example is:

1. 95 percent of medical doctors have bachelors' degrees in biology

2. John Doe is a medical doctor

3. Therefore: John Doe probably has a bachelor's degree in biology

Induction is defined as "inference from particular to general."[4] Inductive logic is closely related to probability theory because it is concerned with the question, "What is the probability that the conclusion is true, given the evidence in question?"[5]

Examples of theories that were developed through an inductive process include Newton's law of gravitation and Boyle's law. They were derived based on observing empirical data and then reasoning to postulate a universal generalization.[6] But data and observations are usually limited. Lipton observes that scientists never have all the relevant data, meaning that reaching a conclusion proven with absolute certainty based on induction is not possible.[7] This is known as underdetermination.

Another complicating factor: often there is more than one hypothesis to explain the observations or clues in view. This is where abductive reasoning enters the picture—allowing us to make an inference to the best explanation. When more than one theory is offered to explain the data, we must then compare how well the competing theories explain the observations. Good examples of inferences to the best explanation are found in criminal investigations, where detectives consider different possible explanations for an unexpected death (e.g., suicide, death by natural causes, murder by a stranger, murder at the hands of someone the victim knew), assemble evidence, and see if one of the explanations stands out as explaining the suite of clues much better than do the other options.

Another example is the investigation of aircraft accidents. In my engineering work on aviation safety, I have examined the findings of a number of aircraft accident investigations. In most accidents there is not conclusive evidence of a single cause; instead there are usually several potential causes. This requires weighing the evidence of the various candidates and determining which potential cause or combination of causes provides the best explanation. Lipton describes this as an inference to "the best of the available potential explanations."[8]

Many scientific theories have been developed through this process of abduction. The existence of the planet Neptune was hypothesized by astronomers John Couch Adams and Urbain Le Verrier based on anomalies in the orbit of Uranus. They used abductive reasoning to infer that the gravitational effects of another planet (later discovered to be Neptune) was the cause.[9]

With abductive reasoning, it's not enough for a theory to explain many of the available clues. Another theory may also explain those clues, and other clues may arise that do not fit the leading explanation, forcing a rethink. (This is practically the plot of every detective show!)

Inductive and abductive reasoning are both commonly applied in science when extrapolating from observations to theories. With induction and abduction, the conclusion cannot be derived with absolute certainty using simple deductive logic.

In the conclusion to *The Origin*, Darwin comments, "It can hardly be supposed that a false theory would explain, in so satisfactory a manner as does the theory of natural selection, the several large classes of facts above specified."[10] There are three problems with his claim. The first is that there are numerous historical examples of theories consistent with the available facts that turned out to be false. One is geocentrism, the idea that Earth is the center of the universe. The known facts were generally consistent with the theory until the observations of Copernicus, and later Galileo and others, along with a mathematically elegant theory of planetary motion offered by Johannes Kepler, together provided a decisive set of clues pointing toward a sun-centered solar system. Another example is luminiferous aether, which was theorized to be the medium that enabled the propagation of light through a vacuum. The theory was considered a valid scientific explanation, consistent with known observations, until experiments eventually proved it false.

The second problem with Darwin's claim is that there are examples of multiple theories consistent with observations, where one is eventually found to be superior.

A third problem with Darwin's comment is that there is another explanation for the origin of life's diversity that arguably explains more of the available evidence, but Darwinians refuse to consider it. That explanation is intelligent design—that life's diversity, including the myriad of complex programmed behaviors reviewed in these pages, was not the product of random mutation and natural selection, or any other blind evolutionary process, but instead was the result of conscious intelligent foresight and engineering. Proponents of the theory appeal only to physical evidence and standard methods of reasoning about past events (e.g., causal adequacy), but evolutionists committed to the rule known as methodological naturalism insist that this competing explanation must not be considered. This move renders evolutionary theory not an inference to the *best explanation* but, less impressively, an inference to the *best allowed explanation*—in this case, the *best purely materialistic explanation*. But if science is a search for truth about the natural world, we should be less interested in the best material explanation than in the best explanation, full stop—meaning intelligent design should be in the conversational mix. Anything less is question begging.

Lipton explains that inferring the best explanation is best accomplished through a careful inventory and comparative analysis, evaluating several factors that are key discriminators. This book has endeavored to do just that, comparing blind evolution and intelligent design as competing explanations for the origin of complex programmed behaviors in animals. Here I want to briefly broaden that exploration to look at blind

METRIC	BLIND EVOLUTION	INTELLIGENT DESIGN
Microevolution	✔	✔
Similarities across taxa	✔	✔
Design flaws	✔	✔
Abrupt appearance		✔
Engineering design		✔
Genetic change		✔
Origin of information		✔
Teleology		✔
Convergence		✔
Simple explanation		✔
Predictions & Retrodictions		✔

Figure 7.1. Comparison of discriminators for two proposed causes for the origin of complex programmed behaviors

evolution and intelligent design more generally, with pointers in the end-notes to additional sources. The overview is necessarily partial in such a quick flyover, and yet it is worth doing in order to show that the evidence against blind evolution explored in the preceding pages of this book, far from unique to CPBs, is part of a larger pattern.

Scientists typically specialize, and a biologist who does this may see problems for evolutionary theory in his or her own subdiscipline but then figure those problems are the exception and that evolutionary theory has things well in hand elsewhere in the life sciences. For this reason it is vital to step back and take in the broader picture. Doing so reveals that while evolutionary theory nicely accounts for some things in the history of life, it runs into significant problems in many other areas. This should give us pause and encourages an objective consideration of how the theory of intelligent design holds up by comparison. Figure 7.1 summarizes such a comparison. A check mark indicates the theory or theories consistent with the observations related to the given category. The remainder of the chapter explains the basis for the comparisons.

Discriminators that Don't

SOME DISCRIMINATORS don't get us very far in deciding which is the better explanation between blind evolution and intelligent design. I've listed these first in the table and will unpack each of them only briefly here.

The first listed is microevolution. One problem with the views of William Paley and other advocates of natural theology prior to Darwin is that they viewed the characteristics of organisms as fixed. Darwin demonstrated in *The Origin* that species were not immutable, but changed over time. A key element of Darwin's evolutionary theory is adaptation. This may involve physical adaptation, such as brown bears evolving into polar bears in arctic regions, or a strain of bacteria developing antibiotic resistance. And it may involve behavioral plasticity as organisms adapt their behavior to changing environments. Here the modern theory of intelligent design has more in common with evolutionary theory than with Paley, since the capacity of plants and animals to adapt to changing environments is understood as fully compatible with a design perspective; after all, an adaptable engineering design is superior to an inflexible one, all other things being equal.

This, by the way, is a case where a systems biology perspective—regarding a biological system as an optimal or near-optimal engineered system for the purpose of studying and better understanding how it works—would lead one to expect and look for system features that enhance adaptability or, as we might say in a purely engineering context, robustness. (More on systems biology below.) Also, much microevolutionary change is devolutionary, and the design perspective has no difficulty with the idea that a design may show some degradation over time. Both perspectives account for the effect of genetic mutations. Even the best engineered systems, after all, show wear and tear over time. So both modern Darwinism and intelligent design well accommodate microevolutionary change. The marvel is that there is so little biological devolution over time, thanks in no small part, we now know, to sophisticated DNA error-correcting mechanisms at the level of genes and proteins.

What about similarities in functional traits (homology) across taxa, such as the recurrence of the pentadactyl (five digit) structure among various animals which is used for such diverse activities as grasping, climbing, crawling, and flying? Evolutionary theory does account for this. Darwin and his followers attribute such common features to inheritance from a common ancestor. At the same time, design also readily accounts for such similarities. Design theorists point to the reuse of successful design strategies in human-made designs, such as the reuse of pulleys, wheels, and gears in widely different technological contexts. (Convergence is a related issue that will be discussed below.)

Another feature of living things that evolutionary theory accounts for is design flaws (dysteleology)—the most obvious being disease and harmful mutations. These are to be expected from a mindless trial-and-error evolutionary process, it is argued. But the design paradigm also accommodates the presence of design flaws. The theory of intelligent design is not a theory of perfect design, and as noted above, even the best designs may degrade over time.

Consider an illustration. If someone who had never encountered modern technology stumbled across a paved road, and in the next moment a man in a convertible Ferrari stopped and gave the primitive fellow a ride, the astonished rider would not be deterred from inferring that the car was the work of intelligent design just because the driver regaled his astonished passenger with the various ways his Ferrari was beginning to give him trouble—underbelly rust, burning too much oil, pulling to the right, etc. That the Ferrari was the product of creative intelligence would remain abundantly obvious to the passenger. In the same way any evidence of imperfect design in the biosphere cannot by itself negate whatever evidence for foresight and conscious design in biology does exist.

Moreover, and as discussed above, those working out of an evolutionary paradigm often are so primed to find bad design in biology that they find it where none exists—as for example in the case of "junk DNA" and so-called "vestigial organs" that turned out to have genuine functions. (More on the issue of imperfect design and dysteleology in Chapter 8.)

In addition to these, there are several other comparisons that can be made between Darwinian evolution and intelligent design. They include overall change over time and the progression from less complex to more complex species. Those issues have not been addressed in the book, but in any event they do not provide a discrimination in favor of either theory, since both models account for these patterns.

Next we move into discriminators that do weigh in favor of one explanation over the other. Taken together these make for a strongly positive case for design.

Abrupt Appearance

A PATTERN of the fossil record well-testified to by mainstream evolutionists is that major new forms appear relatively abruptly. We find this not just in the famous Cambrian explosion, where a little over half a billion years ago all the major animal body plans (phyla) first appeared as sea-going creatures. We find this same pattern at many other points in the history of life. Günter Bechly and Stephen Meyer explain:

> With very few exceptions, the major groups of organisms come into the fossil record abruptly, without discernible connection to earlier (and generally simpler) alleged ancestors in the fossil record. Indeed, leading evolutionary biologists and paleontologists have long acknowledged this pattern of discontinuity. Evolutionary biologist Ernst Mayr, one of the fathers of the modern neo-Darwinian synthesis, famously noted that "wherever we look at the living biota... discontinuities are overwhelmingly frequent.... The discontinuities are even more striking in the fossil record."[11]

Evolutionists have made various attempts to explain this pattern in evolutionary terms, but it remains that the pattern fits much more naturally within a design framework than within an evolutionary paradigm committed to innovation via a series of small, undirected mutations expected to create a branching tree of life, first generating species-level diversity and, only much later, body plan disparity (distinct phyla). Instead the fossil record speaks of abrupt appearance and body plan disparity early on. This pattern flies in the face of evolutionary expectations but fits neatly with intelligent design, since intelligent agents need not, and generally do not, progress through a series of tiny accidental changes in their design work, and designing agents tend to invent fundamentally new forms (the bicycle, the car, the airplane) and then work variations off of those major design categories—technological phyla first, if you will.

As in the case of the abrupt appearance of species, the appearance of CPBs was likely also relatively abrupt. As was demonstrated, CPBs require the coordinated development of several elements, including physiology, behavior algorithms, and neural mechanisms. While some aspects of these may have developed over time, all must occur concurrently and be coherent. That is not consistent with a slow Darwinian process, one random mutation at a time.

Genetic Changes Related to CPBs

RELATED TO the discussion of the abrupt appearance of CPBs, I have argued in these pages that complex programmed behaviors frequently involve the development of or changes to dozens, and even hundreds, of genes. As discussed in Chapter 4, a study of the Asian honey bee (*A. cerana*) genome found 2,182 unique genes, out of a genome consisting of 10,561 genes.[12] Another study compared the genetic changes in seven species of ants, honey bees, and several solitary insects. The study found that each lineage of ants contains about four thousand novel genes, compared to solitary insects.[13] While Darwinian evolution can be the cause of microevolution, the evidence is extremely weak that it can result in the origin of hundreds or even thousands of novel genes, along with novel CPBs, and the information-rich algorithms that appear to underly CPBs.

A common Darwinian explanation for the evolution of novel traits and behaviors is selection pressure. But as discussed in Chapter 6, selection pressure alone cannot explain the origin of novel traits, including the origin of novel CPBs. The better explanation is that these striking innovations in the history of life were intelligently designed, since intelligent agents routinely generate information-rich structures relatively quickly and without the need for their emerging creation to function and improve incrementally at every stage of the creation process.

Engineering Design

THERE HAVE been significant developments recently in the burgeoning discipline of systems biology. According to University of Pittsburgh physicist David Snoke, the "new paradigm" of systems biology "analyzes living systems in terms of systems engineering concepts such as design, information processing, optimization, and other explicitly teleological concepts."[14] Snoke documents how engineering design principles differ from the perspective typical of physics: "The overriding paradigm in physics has been that simple, non-teleological rules will eventually explain everything," he writes. "Even emergent behavior in complex systems is assumed to be the result of simple interactions. By contrast, engineering takes a top-down approach that is explicitly teleological. A goal is defined, and then the parts are arranged to bring about that goal."[15]

In engineering, each system fulfills a specific function. Similarly, in biology each system (digestive, respiratory, nervous, circulatory, etc.) fulfills a specific function. So the leap to using systems engineering thinking in biology is fairly intuitive for engineers.

Systems biology and its use of systems engineering analysis methods can be found in biomimetics and the study of animal communications. Biomimetics is "the study of the formation, structure, or function of biologically produced substances and materials (such as enzymes) and biological mechanisms and processes (such as protein synthesis or photosynthesis) especially for the purpose of synthesizing similar products by artificial mechanisms which mimic natural ones."[16]

The motivation for this field of study is that nature includes numerous designs that appear highly optimized for the given functions. One example is the honey bee nest's honeycomb design, which has now been copied in numerous man-made structures. Another example is silk, which is imitated in materials such as nylon. Other examples are the use of bird wing design to inspire airplane wings, and the design of reduced fluid-drag swimsuits based on shark skin. Also, cell membranes and plant leaves are constructed of materials that are hydrophobic, meaning that they repel water. This was the inspiration for the development of artificial hydrophobic materials used on the hulls of racing boats and antenna radomes (including one such application that I conducted engineering tests on). And as a final example, the air flow design of termite nests described in Chapter 5 was the inspiration for the world's first all-natural cooling building structure in Zimbabwe.[17]

In his fascinating book *Noise Matters*, UNC Chapel Hill biologist R. Haven Wiley applies the engineering field of signal detection theory to animal communications. He states that this application "reveals that the void in our understanding of communication is in fact much greater than naively expected. In particular, the number of parameters required to characterize communications is greater than now realized."[18] Wiley demonstrates that most animal communications, while not achieving "perfect" performance, are nonetheless highly optimized. The reason is that there are always engineering tradeoffs that have to be made among all of the parameters, including the "noise" associated with sensor data.

Based on my own work in communications systems and signal detection engineering, I can confirm this conclusion. In some areas of systems biology, sophisticated mathematical modeling methods are being employed to analyze how biological systems are designed to operate. Such modeling techniques include control systems theory, optimization, information theory, and Bayesian statistics inference.[19] These methods fall under the general category of reverse engineering.

Philosopher of biology Robert Richardson describes how the process of reverse engineering is sometimes used to evaluate the adaptation of biological systems: "If adaptive thinking begins with the ecological 'problems' an organism confronts and explains or infers the 'solution' based on the problem, reverse engineering turns the reasoning around, beginning with the 'solution' and inferring what the ecological problem must have been."[20] According to Richardson the purpose of the reverse engineering in such cases is to infer the historical cause of the physical structure, namely some ecological challenge that evolution then met through adaptation. But Richardson's framing implies that an organism evolves an adaptation to "solve" the "problem." Both words are placed within scare quotes because neither term is justified in the context of Darwinian evolution. Both terms imply goals and purposes, which blind evolution lacks.

On the other hand, goals and purposes are directly applicable from an ID perspective. And on the ID view, reverse engineering in biology works so well because the biological systems we are learning from were indeed engineered—that is, were intelligently designed. An ID perspective, for example, led design theorist William Dembski to correctly predict that much of what Darwinists regarded as "junk DNA" would turn out to have function, and this prediction proved correct.[21]

To summarize: systems biology studies biological systems as if they are optimal or near-optimal engineered systems in order to better understand those systems. Systems biology is practiced widely and is now a fruitful approach in research biology. Many who practice it adhere to evolutionary theory, but while doing systems biology they essentially pretend that they believe the biological system under review was the product of engineering design. Proponents of intelligent design take a less tortured approach to their systems biology work, appealing to the old adage that if it walks like a duck, quacks like a duck, and flies like a duck, then it's probably a duck.

How does systems biology apply to complex programmed animal behaviors? As one example, Chapter 2 described programmed systems for animal navigation in engineering terms, and briefly diagrammed it in Figure 2.2. We find in these animal systems the hardware, sensors, and software we find in many electronic systems. The brain and body parts (legs, wings, etc.) are the hardware. The animals' five senses are the sensors. And the algorithms encoded in the brain are the software. It is also important to reiterate the relationship between form and behavior.

The behavior needs to match the physical structures (form) to achieve optimum performance. That is why in engineering design the process includes evaluating how the hardware can be optimized to perform a function that is coded in the system software. All this maps beautifully onto an analysis of complex programmed animal behaviors. Each of the elements in Figure 2.2 that contributes to the behavior fulfills a specific function. The behavior could not occur unless each of the functions was present and performing in the manner designed. The combination of functions together achieve the goal of the behavior. (See Figure 3.5)

There is also evidence of engineering in the efficiency with which the programs are instantiated in the brains of animals. Social insects are the best example, as they have tiny brains yet contain complex programs governing sophisticated behaviors. The optimization required to embed the algorithms in such small brains is best explained as the product of skillful engineering design.

Eventually the study of these complex programmed behaviors and their underlying architecture may transition to a focus on biomimetics in pursuit of the implementation of compact neural networks. Computer scientists already have begun to explore how to move information storage methods from silicon chips to carbon-based storage,[22] taking a cue from DNA in pursuit of information storage devices that promise to vastly exceed the limits of silicon storage.[23]

How does evolutionary theory fare on the engineering-design discriminator? As we saw above, evolutionists appeal to laws that purportedly require the emergence of this or that cleverly engineered feature. However, the great variety of biological solutions argue against deterministic solutions driven inexorably by natural laws. Finally and most fundamentally, we lack uniform and repeated experience of blind evolutionary processes engineering new systems, whereas we find intelligent agents doing this all the time.

The Origin of Algorithms

As DISCUSSED in Chapter 6, blind evolutionary processes do not explain the origin of the information needed to define complex programmed behaviors, including the algorithms that define and control these behaviors. Blind processes have never been observed to produce truly novel complex and specified information, nor has any such process been properly modeled doing so in a computer environment. Instead, our uniform and repeated experience points to intelligent design as the source of novel functional information, and therefore provides a reasonable explanation for such information in biological systems generally, and for the information underlying CPBs specifically.

An issue not addressed in the studies of genetic changes associated with social behavior and other CPBs is that such changes do not appear to account for the origin of the algorithms that control the behaviors. One possible reason is that the algorithms may not always reside within the protein coding DNA, as there is evidence of changes in non-coding DNA associated with social behavior.[24] Regardless of the exact mechanism, the complexity of the algorithms presents a major challenge to a Darwinian explanation. Conversely, the algorithms exhibit a number of characteristics consistent with intelligent design.

Teleology

TELEOLOGY IS "the doctrine or study of ends or final causes, especially as related to the evidence of design or purpose in nature."[25] Darwinian materialists allow for what is called "internal" teleology—that is, internal to the animal in question. So, according to philosopher Michael Ruse, Darwin "showed how to get purpose without directly invoking a designer—natural selection gets things done according to blind law without making direct mention of mind. The teleology is internal."[26] Francisco Ayala puts it more concretely, "The wings of birds have a natural teleology; they serve an end—flying—but their configuration is not due to the conscious design of any agent."[27] Jerry Coyne is brief and blunt: "The notions of ultimate purpose and 'teleology' (an external force directing evolution) are simply not part of science."[28]

Such an aversion is due, I would argue, to teleophobia, meaning an aversion or unwillingness to admit the existence of design or final causes in nature,[29] since they fit uneasily within a naturalistic paradigm. In addition to asserting that teleology has no role in science, Coyne claims that there is "simply no evidence for the claimed intervention of a teleological designer in evolution."[30] But the evidence is strong that the behaviors described herein (e.g., navigation, migration, architecture) as well as many others do indeed involve teleology, and not just internal teleology as described by Ruse. For example, the social behaviors associated with the eusociality of superorganisms provide evidence for a higher teleology. All of the functions that are performed by social groups have specific purposes and goals that benefit the colony as a whole, including cases where the individual member's chances of surviving and passing on his or her genes is greatly reduced, in contradiction to Darwinian expectations.

Such examples of higher-level teleology are a poor fit with Darwinian materialism but are very much at home under a design perspective. After all, it is precisely agents who can and do plan and pursue goals, both for themselves and others.

Convergence

As we saw in Chapter 6, several similar complex behaviors occur in animals that are widely separated from each other phylogenetically. One example of navigation system convergence is the various somewhat distantly related animals that employ the sun as a compass, including butterflies, ants, bees, and birds. Another example: many types of ants, bees, and termites exhibit social behavior convergence, including caste structures, behavior regulation through pheromones, and nest structure complexity. In addition, ants and termites also exhibit similar farming behaviors.

Evolutionists describe these as the result of convergence. Convergence is simply a label for the phenomenon, and does not actually explain anything. It is often attributed to common selection pressures; however, selection pressure alone cannot be a valid cause. Among other problems, for selection pressure to result in nearly identical novel traits and behaviors in these different animal groups, there would have to be a deterministic process that causes the genetic or epigenetic changes resulting in the behavior. There is no evidence that such a deterministic process can or does occur, and much evidence that it does not.

Intelligent design provides a more reasonable explanation for convergent behaviors: common design by an intelligent agent or agents who can freely choose to reuse ideas and designs in different contexts, much as engineers reuse wheels and other devices in widely divergent machines, or computer programmers reuse software modules in various computer programs.

Simplicity

ONE OF the principles often invoked in scientific explanations is that, other things being equal, the simpler explanation is the one more likely to be correct. This is also known as Occam's razor. Occam's razor would seem to smile on Darwinism. Indeed, simplicity was long one of its main selling points. As Darwin put it in the last sentence of *On the Origin of Species*, "There is grandeur in this view of life, with its several powers, having been originally breathed into a few forms or into one; and that, whilst this planet has gone cycling on according to the fixed law of gravity, from so simple a beginning endless forms most beautiful and most wonderful have been, and are being, evolved."[31] All forms of life, in other words, are said to be due to random variation and natural selection over long periods of time. How simple and elegant!

The appeal to simplicity also has been invoked in discussions about the origin of animal behavior. Psychologist C. Lloyd Morgan invoked a form of Occam's razor known as Morgan's Canon when he proposed that the simplest explanations for the origin of particular animal behaviors are preferred. In sync with this principle, animals—and particularly, less developed animals—were viewed as little more than machines. Examples where this sort of explanation has been invoked include habituation and Pavlovian conditioning. Such behaviors were seen as little more than simple reflexive responses to stimuli.

If this were the end of the story, Occam's razor could reasonably be said to count in favor of blind evolution as an explanation for the origin of CPBs. But as we have seen, there is more to the story. Simplicity can only be a guide to theory virtue if the all-important qualifier "all other things being equal" is kept in view. When it drops away, simplicity becomes a misleading lodestar. And research documented in these pages shows that all other things are far from equal. Simple explanations for the origin of CPBs are insufficient. We saw this in Chapter 4's section on invertebrate learning, where it was shown that insects do not possess a general learning capability, but instead are designed to learn information specific to their lifestyle and behavior. In addition, in Chapter 6 we saw how implementing even simple behaviors involves complex neural mechanisms. We now know that many programmed behaviors are enormously complex. These behaviors involve sophisticated mechanisms, including in some cases sophisticated algorithms, such as those for navigation and social behaviors. Pretending otherwise is to ignore the very thing that needs explaining. And having acknowledged the functional complexity that needs explaining, the next step is to find an explanation that is causally adequate.

If we have one explanation—namely, unguided evolution—that lacks the demonstrated power to produce the complex algorithms underlying complex programmed behavior, and we have another explanation based on a type of cause that has repeatedly demonstrated the power to produce complex algorithms, then all other things are far from equal. The mindless cause has failed an important test, and the design hypothesis has passed it.

On a side note, modern evolutionary theory has even lost much of its simplicity as those seeking to rescue it have festooned it with all manner of add-ons, in a process of accrual not unlike the attempts centuries earlier to rescue the geocentric model of the solar system by adding various epicycles to explain aberrations in the motions of the planets. This accrual process began with subsequent editions of Darwin's *Origin* and has accelerated in recent decades. As we saw in Chapter 1, Darwin included in his theory a form of Lamarckism to explain animal behaviors. So it was actually never as simple as claimed. More recently, the extended evolutionary synthesis sacrifices theoretical elegance while still failing to explain the origin of novel traits, including novel CPBs, and more broadly, the origin of new biological information in the history of life.

Predictions

THIS DISCRIMINATOR is actually a suite of discriminators, and includes some discriminators already touched on above. Blind evolution and intelligent design each lead to certain expectations about what investigators will find in biology and in the history of life. The question is, which model scores the most successes in this area. The jury is still out on some predictions, but for others, the jury returned long ago.

This book is focused on origins biology, so it isn't primarily about making predictions about future events, such as when another solar eclipse will occur. However, scientists talk about making predictions in a second sense—namely, predicting what will or won't be discovered about certain features of the natural world (known as retrodictions). For example, design theorists predicted that much so-called "junk DNA" would be found to have function, and that prediction has proven correct. As a theory of origins, the theory of intelligent design in biology offers more of this second kind of prediction, but it also offers some of the first kind of prediction. For instance, in 2007 Michael Behe commented regarding Richard Lenski's long-term evolution experiment on *E. coli* bacteria:

> The lab bacteria performed much like the wild pathogens: a host of incoherent changes have slightly altered pre-existing systems. Nothing fundamentally new has been produced. No new protein-protein

interactions, no new molecular machines. Like thalassemia in humans, some large evolutionary advantages have been conferred by breaking things…. Breaking some genes and turning others off, however, won't make much of anything. After a while, beneficial changes from the experiment petered out. The fact that malaria, with a billion-fold more chances, gave a very similar pattern to the more modest studies on *E. coli*, strongly suggests that that's all Darwinism can do.[32]

Fourteen years and tens of thousands of *E. coli* generations later, Behe's assessment looks spot on. The *E. coli* bacteria have broken some things that create niche advantages, but nothing fundamentally new has evolved.

As for the subject of the present book, an ID framework for the origin of complex programmed behaviors does suggest some predictions about what we are likely to discover going forward. One prediction implicit throughout these pages is that evidence will continue to mount that CPBs are controlled through complex algorithms and not through "simple" mechanisms as many evolutionists assert. Based on the accumulated evidence, this discriminator already counts in favor of intelligent design over blind evolution. What can we expect going forward? My view is that the evidence for underlying algorithms is already decisive, but this is not a universal view and we have much more to learn about how genes, genetic control elements, epigenetic components, and neural networks combine to generate complex programmed behaviors. I predict that the more we learn the stronger the case for underlying algorithms will be.

Summary of Discriminators

Table 7.1 above lists eleven discriminators—observations that might count in favor of either blind evolution or intelligent design. The table isn't exhaustive, and there is much more to be said about each of the discriminators discussed in the material below the table. The purpose of the discussion was, yes, to make a case for intelligent design over blind evolution as the best explanation for the origin of living forms generally and complex programmed behaviors specifically. But more modestly, it is meant to underscore that the challenges facing evolutionary theory are far from few; they crop up all over the place. By the accounting above, three discriminators can be considered draws and eight discriminators favor intelligent design as the best explanation.

One discriminator we didn't include: "Conforms to the rule of methodological naturalism." For some this is the discriminator above all discriminators, and is a favored tool for ruling intelligent design out of court. But as previously discussed, there are various methodologies among the sciences; methodologies have changed over time; and this particular methodological rule amounts to illogical question-begging when what is under investigation is the true cause of the origin of life's great diversity.

We want to discover the true cause, not the most persuasive explanation consistent with naturalism. And the best way to investigate that question is to wrestle with the evidence, no holds barred. More on naturalism in the final chapter, where we turn to various objections against the theory of intelligent design.

8. Answering Common Objections to Intelligent Design

> There is nothing good that does not meet with opposition, and
> it should not be valued any less because it encounters objections.[1]
> — Vincent de Paul

ANYONE WHO TUNES INTO THE EVOLUTION/DESIGN DEBATE FOR any length of time will soon encounter a series of go-to objections that evolutionists like to lodge against intelligent design theory.[2] Let's consider them here in this final chapter and see how they fare when applied to my case for the intelligent design of complex programmed animal behaviors.

The most fundamental argument against intelligent design is that it should be considered not science but a religious view. Typically this argument rests partly on the presupposition of methodological naturalism. The term naturalism generally refers to ontological naturalism. Ontological naturalism is a "theory of the world that excludes the supernatural or spiritual."[3] Ontological naturalism is also often considered synonymous with materialism and physicalism.[4] This view of naturalism was captured succinctly in the assertion by Carl Sagan in the opening segment of the television show Cosmos: "The cosmos is all that is or ever was or ever will be."[5]

Distinct but related to this is methodological naturalism. The stronger form of methodological naturalism asserts that we are not able to obtain legitimate knowledge of anything that cannot be determined from natural science.[6] The first thing to notice is that this is a philosophical assertion. Indeed, all forms of naturalism are philosophical positions.

As the late University of Notre Dame philosopher Ernan McMullin wrote, "Naturalism is not itself a scientific claim; it would require philosophic argument in its support."[7] In other words, you cannot use science by itself to justify methodological naturalism. This means the claim is self-defeating; it doesn't pass its own test of what constitutes legitimate knowledge. (Picture a guy sawing off the limb he's sitting on.)

In a similar vein philosopher Michael Rea, also at Notre Dame, points to a contradiction in the assertions made by many naturalists. On the one hand, naturalism often includes a commitment to science being the final arbiter of knowledge about the world. On the other hand, naturalists reject the notion that developments in science could cause a rejection of naturalism.[8] But on that approach, scientific evidence isn't being treated as the final arbiter of knowledge about the world; a philosophical rule is.

On a side note, naturalism is frequently yoked to reductionism, the view that biological systems are best investigated at the lowest possible level. Typically this means focusing research on molecular and biochemical causes. This frequently results in a failure to observe and account for the holistic behavior of animals. That includes the teleological elements of many animal behaviors, as illustrated by numerous examples in this book.

Peer Review and Intelligent Design

ANOTHER OBJECTION to intelligent design is that intelligent design scientists haven't published ID-related papers in peer-reviewed journals. This claim is demonstrably false. The Discovery Institute has compiled a list of such publications, which totaled over 130 papers as of July 2017.[9] More have appeared since then. Also, in the history of modern science many important scientific arguments have been put forward in book form with minimal peer review rather than in peer-reviewed science journals. Charles Darwin's *On the Origin of Species* is one such case.

The material presented in this book is based on empirical evidence, the same sort of evidence used to defend and evaluate Darwinian evolution. The evidence and arguments here regarding complex programmed animal behaviors fit within any non-question-begging definition of science. If there is anything that violates the true spirit of science and the scientific revolution, it is the behavior of those who shut down open scientific inquiry by rigidly adhering to methodological naturalism. By refusing to consider the evidence for ID, they arbitrarily limit where the scientific evidence is allowed to lead us.

Yes, ID Is Testable

ANOTHER OBJECTION against intelligent design concerns testability. As a National Academies of Science (NAS) document puts it:

> Intelligent design is not a scientific concept because it cannot be empirically tested.... The argument of creationists reverse the scientific process. They begin with an explanation that they are unwilling to alter—that supernatural forces have shaped biological or Earth systems—rejecting the basic requirements of science that hypotheses must be restricted to testable natural explanations. Their beliefs cannot be tested, modified, or rejected by scientific means and thus cannot be a part of the processes of science.[10]

This NAS statement also appeals to methodological naturalism, but since we just considered that objection, let's set that aside here. Concerning the testability of intelligent design, keep in mind that ID, like neo-Darwinism, is a theory of origins and therefore is focused on unrepeatable past events. Both approaches employ abductive reasoning. This method, widely employed in the historical sciences, involves "inferences made about past events or causes based on present clues or facts."[11] We test both theories by comparing them against the evidence, by seeing if any proposed cause or causes have the demonstrated capacity to produce the effects in question (e.g., produce novel information) and by searching for additional evidence that can help us further discern which explanation is best. This was the focus of the previous chapter, where we tested blind evolution and intelligent design against almost a dozen discriminators.

"Since empirical considerations provide grounds for rejecting historical scientific theories or preferring one theory over another," writes philosopher of science Stephen Meyer, "such theories are clearly testable. Like other scientific theories, intelligent design makes claims about the causes of past events, thus making it testable against our knowledge of cause and effect."[12] Thus, ID is testable but like other theories in the historical sciences, the tests do not involve laboratory experiments that allow us to observe the event in question. How did the animals of the Cambrian explosion arise? Even if one could construct an experiment that evolved a Cambrian-type animal in a lab from a primitive precursor, we wouldn't have observed the Cambrian event in question—namely the original appearance of the Cambrian animals somewhat over half a billion years ago. We would have observed something that scientists might argue is illustrative of the process that produced the first Cambrian animals.

The above what-if is purely illustrative, of course. Scientists have not produced basic biological molecules like DNA or RNA under unguided natural conditions, or intelligently engineered even the simplest cell from scratch. The point is, the testing of historical scientific theories is necessarily a different process from the sort of testing done on, say, a new pharmaceutical drug. In the historical sciences the main test is whether the theory accords with present clues, including our knowledge of the cause-and-effect structure of the world.

Daniel Dennett objects that ID proponents have not predicted anything "crisply denied by the reigning theory."[13] But as noted in the previous chapter, ID theorists have made a number of predictions about what will be discovered, predictions frequently at odds with Darwinian claims. For example, it was ID proponents who determined that the *E. coli* in Lenski's Long Term Evolution Experiments did not evolve anything fundamentally new. Michael Behe made this point in 2007 when he noted that ongoing studies of *E. coli* showed that observed changes involve only slight changes to pre-existing systems, never anything fundamentally new—"No new protein-protein interactions, no new molecular machines. Like thalassemia in humans, some large evolutionary advantages have been conferred by breaking things," but "breaking some genes and turning others off... won't make much of anything." Behe further noted that studies of malaria reveal a similar pattern, and then he flatly asserted that all this "strongly suggests that that's all Darwinism can do."[14] In other words, that's all Darwinism has done or will do. The subsequent several years of experimental evidence have borne out that supposition.[15] Lenski's long-term experiment, for instance, still hasn't generated anything fundamentally new.

The late American philosopher Jerry Fodor and Italian cognitive scientist Massimo Piatelli-Palmarini level a related criticism against intelligent design: "ID makes no predictions at all, testable or otherwise; all of its predictions are post hoc. The trouble is that there is no telling in advance what kind of world an intelligent designer might opt for."[16] But this is to argue with a strawman. There are common characteristics of intelligently designed systems that, as we have seen, can be used to make predictions about future findings of those systems. Additionally, since the theory of intelligent design is a comparative inference to the best explanation, it also makes testable predictions about the limits of blind evolution.

Also, and as Meyer points out, evolutionary theory "makes no prediction about the kinds of traits or species that random mutations and natural selection will produce in the future."[17] Instead, it offers explanations of past events, including the fact that the organisms populating the earth have changed over time. Moreover, and as noted earlier, since evolutionary theory is invoked to explain both optimal design (via natural selection) and bad design, as well as both convergence and divergence, it predicts less than some of its proponents might have us believe. When you bet both heads and tails on every coin toss, you don't get credit for the coin coming up heads or tails.

Philosopher of science Peter Lipton distinguishes between prediction and accommodation. Accommodations are where the "scientist constructs a theory to fit the available evidence."[18] Lipton's definition of a successful prediction is when "the theory is constructed and, with the help of auxiliaries, an observable claim is deduced but, unlike a case of accommodation, this takes place before there is any independent reason to believe the claim is true. The claim is then independently verified."[19]

According to Lipton successful theories usually include both accommodations and predictions, but prediction, he argues, has more value than accommodation. One reason is that accommodation is susceptible to being considered a post hoc explanation, what Lipton calls "fudging." Predictions are superior because they are vulnerable to being falsified. I certainly agree with Lipton and others that making successful predictions has more value than simply accounting for prior observations, even if both are perfectly legitimate tasks of a scientific theory.

Both intelligent design and evolutionary theory, it should be noted, offer various accommodations of existing evidence. For instance, the theory of intelligent design accommodates the discovery of sophisticated genetic information in all life forms. Similarly, many Darwinians explain convergent evolution by appealing to selection pressure. This is a post hoc accommodation.

At the same time, and despite the fact that historical scientific theories do not primarily make predictions that can be tested in the laboratory, "they do," as Meyer notes, "sometimes generate discriminating predictions about what we should find in the natural world—predictions that enable scientists to compare them to other historical scientific theories."[20] Testability, he explains, comes in the form of "predictions about what we are likely to find in living systems as we investigate them."[21] This form of testability includes predictions about what additional evidence will be found in the genomes of organisms, but also extends to predictions about what will and won't be found in the fossil record.

Meyer documented several ID predictions in Appendix A of *Signature in the Cell*. One, noted earlier, concerns the function of so-called "junk DNA" in genomes. As discussed in Chapter 6, many geneticists and most advocates of neo-Darwinian evolution claimed that most of the 99 percent non-coding DNA had no function—thus the label "junk DNA." Design proponents predicted that much of this non-coding DNA would prove to have function, and that prediction has been proven correct.

No, Accepting ID Won't Mean the End of Science

SOME ARGUE that accepting intelligent design will mean the end of science as it has been practiced for centuries. Kenneth Miller writes, "The proponents of ID seek nothing less than a true scientific revolution, an uprising of the first order that would do a great deal more than just displace Darwin from our textbooks and curricula. They seek the undoing of four centuries of Western science."[22] Miller further asserts, "To the ID movement the rationalism of the Age of Enlightenment, which gave rise to science as we know it, is the true enemy. Science will first be redefined, and then the 'bankrupt ideologies' of scientific rationalism can be overthrown once and for all."[23] Obviously these would be dire consequences, if they were actually true. But they aren't.

First, ID proponents use rational arguments based on public evidence about the natural world to make the case for intelligent design. This book consists of a series of rational arguments for ID based on empirical scientific information.

Second, many of the founders of modern science saw evidence of intelligent design in the natural world. The rule of methodological naturalism was glommed onto science later. Miller's appeal is, again, to a methodological dogma masquerading as hard-nosed, empirical, fully rational scientific investigation.

Additionally, Miller's dire warning can be turned back against Darwinian evolution. If a behavior or some other trait is found to be optimized and adaptive, the Darwinian materialist may reflexively attribute it to natural selection. The question of precisely how the trait arose frequently isn't asked. Or when an unusual trait exists in distantly related forms, the evolutionist will cite convergent evolution as the explanation. But convergence is simply a descriptive term, not an actual explanation.

When convergence is cited as an explanation, often it is interpreted to mean there is no need for further research. That stifles the scientific process because additional research is needed to determine the true cause of similar traits. In cases where the presence of common genes in distantly related biological forms is attributed to convergence, further inquiry isn't pursued because the explanation is assumed to be selection pressure, or other unguided evolutionary mechanisms. That stops science.

In contrast, design theorists are open to a given feature of the natural world being caused by natural processes, by intelligent design, or by some combination of the two. Design theorists are free to simply follow the evidence. And as we have seen, systems biology, which approaches biological systems as intelligently engineered systems, is proving enormously fruitful. In sum, ID does conflict with scientific naturalism, but not with science.

Yes, Teleology Has a Place in Science

ANOTHER OBJECTION to intelligent design is that it appeals to teleology (purpose), which is considered verboten by many in the natural sciences. Evolutionary biologist Douglas Futuyma explains why many biologists view teleology as out of bounds. He says biology is now based on a modern form of Darwinism, and Darwin's "alternative to intelligent design was design by the completely mindless process of natural selection. This process cannot have a goal, any more than erosion has the goal of forming canyons, for the future cannot cause material events in the present. Thus the concepts of goals or purposes have no place in biology, except in studies of human behavior."[24]

Thomas Huxley, a contemporary of Darwin, argued against teleology in evolutionary theory: "Teleology implies that the organs of every organism are perfect and cannot be improved; the Darwinian theory simply affirms that they work well enough to enable the organism to hold its own against such competitors as it has met with, but admits the possibility of indefinite improvement."[25]

Resistance to teleology is also evident in some engineers working in the field of systems biology, as illustrated by a comment by bioengineers Adam Arkin and John Doyle: "One danger in using the language of engineering to describe the patterns and operations of the evident products of natural selection is that invoking design runs the risk of invoking a designer."[26]

The notion of teleology is closely associated with the concept of function, which also presents problems for the Darwinian view. Ecologist Innes Cuthill comments that "philosophers have rightly been wary of functional approaches in biology, because to ask the question 'What is behavior "x" for?' is teleological: it implies purpose."[27] French biologist Luc-Alain Giraldeau expresses the typical view of the role of function and the elimination of design in biology: "Function is not a question that is usually addressed in science because function implies a design, and the design a designer."[28] So the motivation for trying to avoid discussion of functions and teleology is that these concepts ultimately imply purposeful intelligent design, and a designer.

Philosopher of biology Tim Lewens asserts that many biologists have mistakenly adopted what he calls the "artifact model of nature: they talk of organisms as though they were designed objects."[29] He disagrees with this approach, insisting that "it is a mistake to think that natural selection is a good analogue to the intentions of a designer.[30] Lewens argues that the usual application of functions in the context of artifacts designed by intelligent agents does not apply to biological organisms because natural selection provides a sufficient naturalistic explanation.[31]

The aversion of some biologists and philosophers of science to using the term "function" is curious, since there are uses of the term that do not even necessarily imply teleology. For example, in physics air pressure is a function of temperature. Or in biology, the internal body temperature of cold-blooded animals is a function of the outside air temperature. And in warm-blooded animals, maintaining a constant body temperature (homeostasis) is a function of the thermoregulatory system.[32] The latter is accomplished through various mechanisms such as sweating, shivering, and blood vessel dilation or constriction. This example poses a problem for evolutionists, because the thermoregulatory system practically screams teleology. However, in the neutral sense of the term function, clearly an evenly regulated body temperature is a function of the thermoregulatory system. So why not freely use the term even if one is an evolutionist committed to purging teleology from the biological sciences? The answer may be more a question for psychology. One is reminded of the famous line from Shakespeare's Hamlet: "Methinks the lady doth protest too much." In other words, the lady is protesting her innocence so obsessively that she's probably guilty. In the same way, I can't help but wonder if some evolutionists, in trying to stamp out any use of the term *function* in biology, are evincing a prickly awareness of just how powerful the impression of design is for engineering marvels like the thermoregulatory system.

Not all biologists are so deterred. Chapter 7 showed how functions are now the common subject of discussion in systems biology. Claiming that the term does not apply in biology is to deny the obvious. And changing the definition in biology, such that it can only be the result of natural selection, is a circular argument.

While some evolutionists are allergic to function in biology, they positively run in terror from outright teleology. Psychologist Hayne Reese asserts that teleology has no place in the analysis of animal behavior, writing, "Behavior analysts reject teleology in its classical sense; in behavior analysis, the purposes of behavior cannot be part of an explanation of the behavior because whether they are mentalisms or private events, they are themselves in need of explanation."[33] In other words, the apparently purposive mental states of the creatures must be reduced to something more basic and free of teleology. What is not considered in this line of reasoning is the possibility that animal behaviors could have purposes and goals that are not necessarily consciously articulated or understood in the mind of the animal. The purpose or goal could originate outside of the animal and be instantiated in the animal via complex programming.

An explanation of animal behavior where teleology is eliminated is given in the following example. If a pigeon pecks at a specific red spot to obtain food we would typically say, "The pigeon pecks the red spot in order to obtain access to the grain hopper." However, that sounds teleological and not proper under Darwinian materialism. The way this is reworded into a non-teleological form based on natural selection is, "The pigeon pecks the red spot because in the history of this pigeon, instances of this response class have been followed by access to the grain hopper."[34] So again, just as with function, the Darwinian paradigm does not allow for any role of teleology in biology.

The acclaimed theoretical biologist Mary Jane West-Eberhard comments on the consequences of this point of view. "Zoologists have engaged in such extreme denial of motivation and goal-directed behavior, not to mention animal consciousness and complex intellectual abilities, that until very recently mechanisms for them are not widely sought or even hypothesized," she writes. "At present, this is perhaps the greatest conceptual void in evolutionary ethology."[35] I could not agree more.

In fairness, some evolutionary biologists frankly admit the reality of teleology in animal behavior. "Goal-directed behavior (in the widest sense of this word) is extremely widespread in the organic world; for instance, most activity connected with migration, food-getting, courtship, ontogeny, and all phases of reproduction is characterized by such goal orientation," commented influential evolutionary biologist Ernst Mayr. "The occurrence of goal-directed processes is perhaps the most characteristic feature of the world of living organisms."[36]

Indeed, virtually all categories of animal behaviors include goals, and thus are teleological. Foraging has the goal of finding food. Mating behavior and rearing of offspring have the goal of propagating the species. Migratory behavior has the goal of traveling to a specific destination. Navigation systems have the purpose of enabling accurate migrations. Communication behaviors by social insects have the goal of coordinating the actions of the colony, including nest construction and maintenance. Teleology clearly exists at the level of organismal behavior, and complaints about design theorists allowing it at the level of ultimate explanation spring from the same source as calls to remove teleological language from biology at every level—a philosophical commitment to reductionism and materialism.

Against Fixity of Species

ADVOCATES OF Darwinian evolution sometimes argue that ID is false because if it were true, then each animal would be perfectly designed (by an omnipotent and all-wise God) to function with 100 percent efficiency within its ecological niche. This theologically charged objection assumes, among other things, that all organisms were created at a single moment and were not intended to be adaptable to changing environments.[37] That objection is valid when applied to creationists who held that God had created all of life in a perfect state and that species were unvarying. This view was advocated by William Paley and others. Darwin successfully rebutted this view in *The Origin*. Kenneth Miller, a latter-day Darwin defender, argues along similar lines in his 2008 book *Only a Theory*:

> What does the reality of natural history say with respect to the notion of design? First, it tells us that the specific details of today's living organisms were not the direct product of a flash of design in the dim and distant past. If they had been, life wouldn't be about change; it would be about stasis. Second, it tells us that the ability to adapt to change and even to generate it is one of the most striking characteristics of living things. The design of life, ironically, includes an ability to change its own design.[38]

There are at least two problems here. First, stasis is a pervasive feature of the history of life, a point acknowledged by various mainstream evolutionists, including perhaps most prominently Harvard paleontologist Stephen Jay Gould.[39] Second, Miller is making a strawman argument if he means it to apply to the modern theory of intelligent design. Design theorists happily acknowledge that species vary within a range and often in response to environmental changes. Indeed, one line of design reasoning draws on engineering principles to argue that the capacity of biological forms to adapt is evidence of a form of sophisticated design and is therefore additional evidence for these systems being the work of not just intelligent design but of highly intelligent design. Certainly the approach taken in the present book does not assume that species are fixed. As discussed in Chapter 3, many animal types exhibit plasticity in their behavior.

The Suboptimal Design Objection

IN ANOTHER theologically charged objection, ID critics argue that instances of suboptimal design in biology demonstrate that such systems are not the work of intelligent design but rather came about through blind evolution's trial-and-error process. Mayr describes this as part of Darwin's motivation for introducing adaptation as a "replacement for supernatural design. Design, as conceived by the natural theologians, had to be perfect, for it was unthinkable that God would make something that was less than perfect."[40]

But this imperfection is supposed to flow naturally out of evolution. Daniel Dennett argues that Darwinian evolution, in contrast to an intelligent agent such as an engineer, lacks foresight and therefore "unforeseen or unforeseeable side effects are nothing to it; it proceeds via the profligate process of creating vast numbers of relatively uninsulated designs, most of which are hopelessly flawed because of self-defeating side effects, but a few of which, by dumb luck, are spared that ignominious fate."[41] Francisco Ayala makes an even more extreme form of this argument: "The design of organisms is not intelligent, but rather quite incompatible with the design that we would expect of an intelligent designer or even of a human engineer, and so full of dysfunctions, wastes, and cruelties as to unwarrant its attribution to any being endowed with superior intelligence, wisdom, and benevolence."[42]

But the argument of suboptimal design is logically flawed. First, and as noted, a designed object can show clear evidence of design despite exhibiting imperfections. Something can be less than optimally engineered and still carry the hallmark of having been designed by a conscious agent. Also, it is now widely acknowledged that the biological realm is filled with structures, machines, and systems that vastly outstrip our most advanced technologies. Moreover, many things once deemed suboptimal have turned out to have sound reasons for what was deemed suboptimal, such as the so-called "backward wiring" of the vertebrate eye.[43]

Additionally, the process of designing any object or structure involves trade-offs among several attributes. With biological structures the attributes can include functional efficiency, energy use, complexity, and flexibility. In many cases these attributes conflict, such that if one is optimized others are less than optimal. It is nearly impossible for an individual structure to be optimized for all attributes. The optimal design in many cases therefore represents a compromise among the competing attributes. That is a basic principle of engineering known as constrained optimization. There is no such thing as a "perfect design" in engineering where perfection is understood as a result free of trade-off constraints.

The same principle applies to animal behavior. There is no such thing as trade-off-free behavior. As noted in Chapter 7 in the discussion about animal communication, there is always "noise" associated with the transmission and reception of information. Such noise complicates the process of determining exactly what the information represents. That applies to the information available to an animal relative to all decision making. Because of the noise, the decision likely will not be optimal. However, as we have seen in many areas, animals routinely make near-optimal decisions, such as in the case of insect social behaviors. The same thing is true for architecture, where the design of many nests is near optimal. Another example is the navigation systems of many animals, which have been designed such that the animals achieve near optimal navigation performance, given the accuracy of the available sensory information.

The Problem of "Evil" Animal Behavior

DARWIN AND other evolutionists have used what is known as "the problem of evil" to argue for mindless evolution and against the view that God created life's diversity. A good God, the say, would not have created our biosphere, since it is rife with cruelty, pain, and death. Here is how Darwin put it in a letter to Asa Gray:

> There seems to me too much misery in the world. I cannot persuade myself that a beneficent & omnipotent God would have designedly created the *Ichneumonidae* with the express intention of their feeding within living bodies of caterpillars, or that a cat should play with mice…. I am inclined to look at everything as resulting from designed laws, with the details, whether good or bad, left to the working out of what we may call chance.[44]

In another letter Darwin wrote, "What a book a devil's chaplain might write on the clumsy, wasteful, blundering, low, and horribly cruel works of nature."[45]

Similarly, Richard Dawkins describes the magnitude of animal suffering: "The total amount of suffering per year in the natural world is beyond all decent contemplation. During the minute it takes me to compose this sentence, thousands of animals are being eaten alive; others are running for their lives, whimpering with fear; others are being slowly devoured from within by rasping parasites; thousands of all kinds are dying of starvation, thirst and disease."[46] Dawkins argues that what is a problem for theism disappears under Darwinian materialism, since on the latter view, there is no Creator and the categories of good and evil are mere illusion.

Similarly, philosopher of science and agnostic Michael Ruse contrasts what we observe to the type of creation he believes a good God would have designed: "After Darwin we see that the world is simply not how it would be if an all-loving, all-powerful creator had made it. Nothing like this could possibly occur, given a loving god. But this is exactly what one would expect from a blind, purposeless law."[47]

Examples of morally disturbing animal behavior are abundant. Most female mammals will attack and sometimes kill the young that are not their own. The adaptive explanation for this is that it protects females from wasting care on non-related young and eliminates competition for scarce resources she wants for her own young.[48] If they were to care for young that are not their own it would take away time and resources from caring for their own offspring. Male lions sometimes kill cubs already present when they take over prides of females. This makes sense from the perspective of selfish genes, by maximizing the offspring of the male, in part because it shortens the time for the female to go into estrus again and be capable of having cubs fathered by the new male.[49] Therefore, neo-Darwinian evolution, with its emphasis on natural selection, appears to favor these kinds of behaviors.

In some instances parents engage in cannibalism of offspring, which is somewhat common in fish. Research has shown that this may be an adaptive behavior, by providing an extra source of food for remaining offspring, or by eliminating the brood completely and enabling greater chance of success in the following brood.[50]

Another common phenomenon is siblicide, where one offspring kills another. It is relatively common in birds, but also occurs in some insects and amphibians. A significant difference from birds is that insects and amphibians usually consume the victim for food.[51] There are two forms of siblicide—obligate and facultative. In obligate siblicide two eggs are hatched and one offspring kills the other. We see this with masked booby birds. Two chicks are born several days apart. The one born first is usually stronger and will kill the one born second, by forcing it out of the nest.[52] Facultative siblicide depends upon environmental circumstances, typically the availability of food for the chicks. Blue-footed booby chicks exhibit this behavior. From the point of view of the surviving sibling, it gives it the advantage of more resources without a competing sibling around.

Various plausible adaptive explanations have been offered for siblicide.[53] The explanations are somewhat different in the two forms. The behavior can be viewed as a form of risk management. Although not always related to food resources, it usually occurs when the food provided to the chicks is limited, such that keeping both would risk losing them both. Where food is more abundant, keeping both chicks not only improves the chances of more descendants but also provides insurance in case one of them dies.

Primates are also known to practice infanticide. In the case of Hanuman langur monkeys in India, adult males will kill young infants from females within their social group.[54] Research indicates the behavior tends to occur when males move into a new group, followed by the ejection of the adult male that has fathered the infant langurs. In doing so, the females will become available for becoming pregnant again, presumably by the invading male. Thus the behavior benefits the male by enabling him to produce more offspring. Similar behavior has been found in lions, horses, rodents, and bats.[55]

Another animal behavior that will strike many of us as morally disturbing: brown bears are much more likely to abandon single offspring than twins.[56] This can be explained as an adaptive behavior because the mama bear can then initiate another reproductive cycle, increasing the likelihood of producing multiple offspring that survive.

Joan Strassmann and David Queller describe another example of "evil" animal behavior: "When it comes time for honey bee colonies to divide, several half-sister queens are reared with special food in extra-large cells. The old queen leaves with much of the workforce to start a new colony. Then, the first of the new queens to emerge as an adult seeks out all of the other queen cells and uses her sting to kill her sister rivals."[57]

These examples pose a challenge for those convinced that such creatures are the work of a benevolent designer. In his critical review of Phillip Johnson's book *Darwin on Trial*, David Hull writes, "Whatever the God implied by evolutionary theory and the data of natural history may be like, He is not the Protestant God of waste not, want not. He is also not a loving God who cares about His productions. He is not even the awful God portrayed in the book of Job. The God of the Galápagos is careless, wasteful, indifferent, almost diabolical. He is certainly not the sort of God to whom anyone would be inclined to pray."[58] Hull thus joins Dawkins, Dennett, and others in using apparent "evil behavior" in making an argument not just for unguided evolution but for atheism.

Cornelius Hunter pushes back. In *Darwin's God* he describes how Darwin and others have used the argument from "evil" animal behavior in support of blind evolution. Hunter describes this argument as theological and "metaphysical because it requires certain premises about the nature of God. A great irony reveals itself here: evolution, the theory that made God unnecessary, is itself supported by arguments containing premises about the nature of God."[59] As Hunter explains, Darwin's argument distances God from the natural evil in the world, and his theory of evolution is "very much a solution to the problem of natural evil."[60]

The Darwinian solution comes at a cost, though. "Separating God from creation and its evils meant that God could have no direct influence or control over the world,"[61] Hunter writes. As this Darwinian solution is taken to its logical conclusion, the problem of evil disappears altogether because there is no God, nothing at bottom but matter and energy. The problem of evil disappears because evil itself, along with good, disappears as a real entity. There is left only the appearance of good and the appearance of evil, taking their place alongside the appearance of design in biology in materialism's vison of reality as a great joyless funhouse of illusions.

The general problem of pain and evil in the world has been the subject of much reflection. There does not appear to be a completely satisfactory answer at this time without recourse to certain theological ideas.[62] This should not be surprising, however, since this objection to design is itself theological. The critic of design should not level a theological objection and then complain when a theological response is offered.

An extended treatment of the issue of animal suffering would go beyond the scope of this book. Suffice it to say, we should not forget that the great challenge for any theory of blind evolution is not to explain dysfunction and poor design but to explain the origin of sophisticated functional biological form, machinery, and behavior. At this, as I and others have argued, evolutionary theory has failed, offering only a series of bluffs, just-so stories, and promissory notes in place of cogent evidence of causal adequacy.

As a final aside, the problem of evil behavior presents a noteworthy contrast to the problem of explaining altruistic behavior. As discussed in Chapter 3, certain altruistic social behaviors present a challenge to neo-Darwinian evolution because based on the "selfish gene" principle, animals should not deliberately sacrifice their lives or their ability to reproduce in deference to other members of their colony who do not necessarily share their genes. Conversely, altruistic behavior is consistent with ID, because an intelligent agent can design behaviors that best serve the overall long-term goals of a social group. The question then becomes, Which outlook best explains both sophisticated functional designs and dysfunction or degradation; both selfish behavior and altruistic behavior; and both the reality of a universe governed by indifferent laws as well as the powerful sense that good and evil are real and not mere illusion? Which paradigm is rich enough to accommodate all of this?

Who Designed the Designer?

RICHARD DAWKINS poses another objection to intelligent design. Here is how he summarizes it in *The God Delusion*:

> 1. One of the greatest challenges to the human intellect over the centuries has been to explain how the complex, improbable appearance of design in the universe arises. 2. The natural temptation is to attribute this appearance of design to actual design itself.... 3. The temptation is a false one, because the designer hypothesis immediately raises the larger problem of who designed the designer.[63]

Dennett concurs. "If God created and designed all these wonderful things, who created God?" he asks. "Supergod? And who created Supergod? Or did God create himself? We may ask instead whether this bland embrace of mystery is any improvement over just denying the principle that intelligence (or design) must spring from intelligence."[64]

Note that this argument is a philosophical one, and not scientific. Moreover, there are significant problems with it, regardless of how one labels the argument. One problem is that the theory of intelligent design does not posit that "intelligence... must spring from intelligence." This is a mischaracterization. As a theory of origins intelligent design is con-

204 / ANIMAL ALGORITHMS /

cerned only with things that came into being, that originated. It doesn't apply to anything that has always existed; even less to something that, by definition, exists necessarily from all eternity. The God of theism is understood as such a being. So for the theist convinced that God is the source of the intelligent design evident in the natural world, the answer to "who designed the designer" is simple: No one. The Designer has always existed.

Philosopher William Lane Craig explains another problem with the who-designed-the-designer objection: "In order to recognize an explanation as the best, one needn't be able to explain the explanation. In fact, so requiring would lead to an infinite regress of explanations, so that nothing could ever be explained and science would be destroyed! So in the case at hand, in order to recognize that intelligent design is the best explanation of the appearance of design in the universe, one needn't be able to explain the Designer."[65] The book *Who Designed the Designer?* by Michael Augros provides a detailed philosophical exposition of why a further explanation of the designing intelligence is not required.[66]

Not God-of-the-Gaps or an Argument from Ignorance

SOME WILL assert that the argument presented here is simply a God-of-the-gaps argument, according to which if science remains ignorant of precisely how nature brought about something in biology, God, or at least some intelligent designer, must have done it. If that was the structure of the contemporary argument to intelligent design, it would indeed be fallacious. But that isn't in fact the structure of the argument. The argument is not simply based on the absence of evidence for a Darwinian explanation. Rather it is a positive argument for design. Chapter 7 summarized the positive evidence supporting ID.

The more general form of the "God-of-the-gaps" charge is known as an "argument from ignorance." According to this fallacious mode of reasoning, Theory A is not true; therefore, Theory B must be true. But just because one theory about some phenomenon has been proven to be false does not mean that a proposed alternative theory is therefore true. In this case, the criticism is that ID claims that a particular biological characteristic could not be the result of neo-Darwinian evolution; therefore ID must be the explanation—a fallacious argument from ignorance. But again, this mischaracterizes the case for intelligent design.

It is true that part of the contemporary design argument in biology is that neo-Darwinian evolution cannot produce a particular biological characteristic, such as the bacterial flagellum or the structure and coding of DNA. However, the design argument also makes a case against other mindless evolutionary scenarios and offers positive evidence for intelligent design. Contemporary design theorists make an argument to the best explanation, the abductive mode of reasoning common to the historical sciences. That holds for the case I made in these pages as well. It involves both negative arguments against competitor explanations and positive evidence for design, including evidence for teleology in complex programmed animal behaviors, and evidence for the engineering design of many CPBs. The argument is founded not on ignorance but on knowledge of the cause-and-effect structure of the world.

Final Thoughts

THIS BOOK does not, of course, provide a complete explanation for the origin of complex programmed animal behaviors. I have argued more modestly that intelligent design is the best explanation for CPBs, involving as it does a cause with the demonstrated power to produce the underlying algorithms and other information required of CPBs, while mindless evolution appears to lack any such demonstrated power. Rather than answer all questions, the design hypothesis leaves open many questions regarding the specific mechanisms through which the designing intelligence instantiated the CPBs. Much more remains to be discovered using a systems biology approach to CPBs—that is, regarding the systems as instances of near optimal engineering and then employing reverse engineering to better understand how they work. This includes discovering much more about the specific form the programming and coding takes in the animal brains, still very much a mystery.

Thus the design inference is not the end of science, as claimed by opponents of ID. Rather, it opens the door to a wider range of scientific investigation.

ENDNOTES

1. GENIUS IN LILLIPUT

1. Mary Jane West-Eberhard, *Developmental Plasticity and Evolution* (Oxford: Oxford University Press, 2003), 314.

2. Frederic Libersat, "Wasp Uses Venom Cocktail to Manipulate the Behavior of its Cockroach Prey," *Journal of Comparative Physiology A* (2003) 189: 497–508, https://doi.org/10.1007/s00359-003-0432-0.

3. Katsuhiro Konno, Kohei Kazuma, and Ken-ichi Nihei, "Peptide Toxins in Solitary Wasp Venoms," *Toxins* 8, no. 4 (April 18, 2016): 114, https://doi.org/10.3390/toxins8040114.

4. James L. Gould and Carol Grant Gould, *Animal Architects* (New York: Basic Books, 2007), 16.

5. Jerry Fodor and Massimo Piattelli-Palmarini, *What Darwin Got Wrong* (New York: Farrar, Straus and Giroux, 2010), 91.

6. Jennifer Ackerman, *The Genius of Birds* (New York: Penguin Press, 2016).

7. Martin Giurfa, "The Amazing Mini-Brain: Lessons from a Honey Bee," *Bee World* 84, no. 1 (2003): 5–18.

8. Robin Baker, *The Mystery of Migration* (New York: Viking Press, 1981).

9. James L. Gould and Carol Grant Gould, *Nature's Compass: The Mystery of Animal Navigation* (Princeton: Princeton University Press, 2012).

10. Charles Darwin, *On the Origin of Species: By Means of Natural Selection, or The Preservation of Favoured Races in the Struggle for Life* (London: John Murray, 1859), 207, https://archive.org/details/onoriginspeciesf00darw/page/n9/mode/2up.

11. Darwin, *On the Origin of Species*, 209–210. *Natura non facit saltum* is Latin for "Nature does not make jumps."

12. Ernst Mayr, *Toward a New Philosophy of Biology* (Cambridge, MA: Harvard University Press, 1988), 408.

13. Ernst Mayr, *What Evolution Is* (New York: Basic Books, 2001), 137.

14. Aristotle, *The Complete Works of Aristotle* (United Kingdom: Delphi Classics, 2013), loc. 27355, Kindle.

15. Jean-Baptiste Lamarck, *Zoological Philosophy: An Exposition with Regard to the Natural History of Animals* [1914] (London: Forgotten Books, 2012), 107.

16. Lamarck, *Zoological Philosophy*, 113–114.

17. Lamarck, *Zoological Philosophy*, 120.

18. Lamarck, *Zoological Philosophy*, 122.

19. Peter Bowler, *Evolution: The History of an Idea* (Berkeley: University of California Press, 1983), 160.

20. Darwin, *On the Origin of Species*, 11. Darwin writes, "Not a single domestic animal can be named which has not in some country drooping ears; and the view suggested by some authors, that the drooping is due to the disuse of the muscles of the ear, from the animals not being much alarmed by danger, seems probable."

21. Darwin, *On the Origin of Species*, 134.

22. Darwin, *On the Origin of Species*, 143.

23. Darwin, *On the Origin of Species*, 207.

24. Darwin, *On the Origin of Species*, 243.

25. Darwin, *On the Origin of Species*, 216.

26. Darwin, *On the Origin of Species*, 209.

27. Charles Darwin, *The Descent of Man and Selection in Relation to Sex* [1871] (Amherst, NY: Prometheus Books, 1998), 70.

28. Bowler, *Evolution: The History of an Idea*, 337.

29. Bowler, *Evolution: The History of an Idea*, 251.

30. Michael Lynch, "The Frailty of Adaptive Hypotheses for the Origins of Organismal Complexity," in *In the Light of Evolution*, vol. 1, *Adaptation and Complex Design*, eds. John C. Avise and Francisco J. Ayala (Washington: National Academies Press, 2007), 87.

31. Lynch, "The Frailty of Adaptive Hypotheses," 86.

32. As Richard Burkhardt explains, "Where the ethologists were interested primarily in the areas of behavioral evolution, causation, and function, the comparative psychologists were more interested in behavioral development." Richard W. Burkhardt, Jr., *Patterns of Behavior: Konrad Lorenz, Niko Tinbergen, and the Founding of Ethology* (Chicago: University of Chicago Press, 2005), 384.

33. Burkhardt, *Patterns of Behavior*, 361.

34. Burkhardt, *Patterns of Behavior*, 202. This is a summary by Burkhardt of the paper presented by Lorenz at a symposium in Leiden, "Uber den Begriff der Instinkhandlung." The title of his address at the symposium was "On the Concept of Instinct in Animal Psychology." The idea that behavior patterns could be used to reconstruct phylogenetic trees was not novel to Konrad Lorenz. It had previously been proposed by the German Oscar Heinroth and the American Charles Otis Whitman. See Burkhardt, *Patterns of Behavior*, 134.

35. Niko Tinbergen, *The Study of Instinct* (New York: Oxford University Press, 1974).

36. Niko Tinbergen, "On Aims and Methods of Ethology," *Zeitschrift für Tierpsychologie* 20 (1963): 410–433, https://doi.org/10.1111/j.1439-0310.1963.tb01161.x. Reprinted in *Tinbergen's Legacy: Function and Mechanisms in Behavioral Biology*, eds. Johan Bolhuis and Simon Verhulst (New York: Cambridge University Press, 2009), 1–24.

37. Patrick Bateson and Peter Klopfer capture the sea change, if not all the nuances, in their comment that "ethology as a coherent body of theory ceased to exist in the 1950s." P. P. G. Bateson and Peter H. Klopfer, preface to *Perspectives in Ethology*, vol. 8, *Which Ethology?*, eds. P. P. G. Bateson and Peter H. Klopfer (New York: Plenum Press, 1989), v.

38. Mauricio Papini, *Comparative Psychology: Evolution and Development of Behavior* (New York: Psychology Press, 2008), 7. Pereira defined three types of movements by animals. One is instincts produced by sensory experience, such as chicks pecking soon after hatching. The second is movements that occur without any external stimulus. The third type involves learning through repetition, such as a parrot repeating speech. He believed that only humans were capable of conscious thought.

39. John Alcock, *Animal Behavior: An Evolutionary Approach* (Sunderland, MA: Sinauer Associates, 2013), 362.

40. James L. Gould and Carol Grant Gould, *Animal Minds* (New York: Scientific American Library, 1994), 24.

41. Konrad Lorenz, "Betrachtungen uber das Erkennen der arteigenen Triebhandlungen der Vogel," *Journal für Ornithologie* 80 (1932): 50–98, quoted in Burkhardt, *Patterns of Behavior,* 146.

42. Tinbergen, "On Aims and Methods of Ethology," in *Tinbergen's Legacy,* 5. Tinbergen also clarified that this ignores variations due to differences in environment during ontogeny.

43. Mark S. Blumberg, *Basic Instinct: The Genesis of Behavior* (New York: Thunder's Mouth Press, 2005), xiv.

44. Patrick Bateson, foreword to *The Development of Animal Behavior: A Reader,* eds. Johan J. Bolhuis and Jerry A. Hogan (Oxford: Blackwell Publishers, 1999), x.

45. Sara Shettleworth, *Cognition, Evolution, and Behavior* (Oxford: Oxford University Press, 2010), 14. The same point is explained in P. Bateson and M. Mameli, "The Innate and the Acquired: Useful Clusters or a Residual Distinction from Folk Biology?," *Developmental Psychobiology* 49 (2007): 818–831.

46. Shettleworth, *Cognition, Evolution, and Behavior,* 13.

47. Jerry A. Hogan and Johan J. Bolhuis, "The Development of Behavior: Trends since Tinbergen," in *Tinbergen's Legacy,* 104.

48. Blumberg, *Basic Instinct,* 17.

49. Blumberg, *Basic Instinct,* 17.

50. J. Scott Turner, *Purpose and Desire: What Makes Something "Alive" and Why Modern Darwinism Has Failed to Explain It* (New York: HarperOne, 2017), 8.

51. Robert K. Barnhart, *Hammond Barnhart Dictionary of Science* (Maplewood, NJ: Hammond, 1986), s.v. "algorithm."

52. David Berlinski, *The Advent of the Algorithm: The Idea That Rules the World* (New York: Harcourt, 2000).

53. Mayr, *Toward a New Philosophy of Biology,* 49.

54. Shettleworth, *Cognition, Evolution, and Behavior,* 18.

55. *Principles of Neural Science,* eds. Eric Kandel, James Schwartz, and Thomas Jessell, 4th ed. (New York: McGraw-Hill, 2000), 42.

56. Jay C. Dunlap, "Molecular Bases for Circadian Clocks," *Cell* 96 (January 22, 1999): 271–290.

57. Z. Yao and O. T. Shafer, "The *Drosophila* Circadian Clock Is a Variably Coupled Network of Multiple Peptidergic Units," *Science* 343, no. 6178 (March 28, 2014): 1516–1520, https://doi.org/ 10.1126/science.1251285.

58. Yehuda Ben-Sharar, "Epigenetic Switch Turns on Gene Behavioral Variations," *PNAS* 114, no. 47 (November 21, 2017): 12365–12367, https://doi.org/10.1073/pnas.1717376114.

59. Changwoo Seo et al., "Intense Threat Switches Dorsal Raphe Serotonin Neurons to a Paradoxical Operational Mode," *Science* 363, no. 6426 (February 1, 2019): 538–541. The two neuron types are dorsal raphe nucleus (DRN) 5-HT and dorsal raphe g-aminobutyric acid (GABA) neurons.

60. Quantitative genetics is concerned with the evaluation of variable traits that are influenced by multiple genes.

61. Massimo Pigliucci and Gerd B. Müller, *Elements of an Extended Evolutionary Synthesis* (Cambridge, MA: MIT Press, 2010), 14.

62. Scott P. Carroll and Patrice Showers Corneli, "The Evolution of Behavioral Norms of Reaction as a Problem in Ecological Genetics," in *Geographic Variation in Behavior: Perspectives on Evolutionary Mechanisms*, eds. Susan A. Foster and John A. Endler (New York: Oxford University Press, 1999), 52. Carroll and Corneli go on to describe how the behavioral differences among individual animals can have three different sources: (1) The environmental conditions they experience may vary (2) The genes that code for specific tactics or predispositions may differ (3) The genotype/environment interaction, manifested through developmental pathways, may vary.

63. David F. Westneat and Charles W. Fox, eds., *Evolutionary Behavioral Ecology* (New York: Oxford University Press, 2010), 127.

2. NAVIGATION AND MIGRATION

1. Lucius Annaeus Seneca, *Letters from a Stoic*, trans. Richard M. Gummere (Digireads.com Publishing, 2017), loc. 180, Kindle.

2. James L. Gould and Carol Grant Gould, *Nature's Compass: The Mystery of Animal Navigation* (Princeton: Princeton University Press, 2012), 3.

3. Ben Hoare, *Animal Migration: Remarkable Journeys in the Wild* (Berkeley: University of California Press, 2009), 143.

4. Gould and Gould, *Nature's Compass*, 16.

5. Gould and Gould, *Nature's Compass*, 174.

6. James L. Gould and Carol Grant Gould, *Animal Architects* (New York: Basic Books, 2007), 17.

7. Gould and Gould, *Nature's Compass*, 176.

8. Rüdiger Wehner et al., "Ant Navigation: One-Way Routes Rather than Maps," *Current Biology* 16 (January 10, 2006): 75–79, https://doi.org/10.1016/j.cub.2005.11.035.

9. Donald Launer, *Navigation Through the Ages* (Dobbs Ferry, NY: Sheridan House, 2009), 21. Inertial navigation and path integration are more complex versions of dead reckoning.

10. Gyroscopes measure angular velocity, while accelerometers measure linear acceleration of aircraft in three dimensions.

11. Wehner et al., "Ant Navigation: One-Way Routes Rather than Maps," 75.

12. Gould and Gould, *Nature's Compass*, 167.

13. Gould and Gould, *Nature's Compass*, 31.

14. Thomas Stone et al., "An Anatomically Constrained Model for Path Integration in the Bee Brain," *Current Biology* 27 (October 23, 2017): 3069–3085, https://doi.org/10.1016/j.cub.2017.08.052.

15. Stone et al., "An Anatomically Constrained Model," Simulation Methods section.

16. Launer, *Navigation Through the Ages*, 24.

17. Launer, *Navigation Through the Ages*, 24.

18. Launer, *Navigation Through the Ages*, 37. To make it even more complicated, the North and South poles reverse periodically. This has occurred twenty-five times in the last five

million years. In addition, during the changeover the magnetic field becomes zero. The last reversal occurred approximately 780,000 years ago.

19. Kenneth J. Lohmann, "Magnetic-Field Perception," *Nature* 464 (April 22, 2010): 1140.

20. Wolfgang Wiltschko and Roswitha Wiltschko, "Magnetic Orientation in Birds," *Journal of Experimental Biology* 199, no. 1 (January 1996): 31.

21. Wiltschko and Wiltschko, "Magnetic Orientation in Birds," 31.

22. Robert C. Beason, "Mechanisms of Magnetic Orientation in Birds," *Integrative Comparative Biology* 45, no. 3 (June 2005): 565–573, https://doi.org/10.1093/icb/45.3.565.

23. Gould and Gould, *Nature's Compass*, 114.

24. Gould and Gould, *Nature's Compass*, 212–213.

25. James L. Gould, "Magnetoreception," *Current Biology* 20, no. 10 (2010): R435.

26. The magnetite that has been found in most animals appears as what are called single-domain crystals that are extremely small (about 50 micrometers). These are permanent magnets that align with the Earth's magnetic field (Lohmann, "Magnetic-Field Perception.") There have been some limited experiments that suggest humans might have a magnetic sense. The strongest evidence has come from experiments conducted by Caltech researcher Joe Kirschvink that have yielded positive indications of a magnetic sense. Kirschvink's research is described in Eric Hand, "Polar Explorer," *Science* 352, no. 6293 (June 24, 2016): 1509–1513. One fact that supports the possibility of human magnetic sense is that magnetite has been found in the human brain (Stuart A. Gilder et al., "Distribution of Magnetic Remanence Carriers in the Human Brain," *Nature: Scientific Reports* 8 (July 27, 2018): 11363, https://doi.org/10.1038/s41598-018-29766-z). However, despite some evidence humans may possess a magnetic sense, the Goulds have been skeptical. "In common with our fellow primates, there is no good reason to believe that we have either a map sense or a magnetic compass," they write. (Gould and Gould, *Nature's Compass*, 224–225).

27. Eric Hand, "What and Where Are the Body's Magnetometers?," *Science* 352, no. 6293 (June 24, 2016): 1510–11.

28. Roswitha Wiltschko and Wolfgang Wiltschko, "Sensing Magnetic Directions in Birds: Radical Pair Processes Involving Cryptochrome," *Biosensors* 4 (July 24, 2014): 222.

29. Le-Qing Wu and J. David Dickman, "Neural Correlates of a Magnetic Sense," *Science* 336 (May 25, 2012): 1054–1057.

30. Gould and Gould, *Nature's Compass*, 19.

31. Talbot Waterman, *Animal Navigation* (New York: Scientific American Library, 1989), 100.

32. Gould and Gould, *Nature's Compass*, 87–90.

33. Gould and Gould, *Nature's Compass*, 73.

34. Rüdiger Wehner, "The Desert Ant's Navigational Toolkit: Procedural Rather Than Positional Knowledge," *Journal of the Institute of Navigation* 55, no. 2 (Summer 2008): 109.

35. Rüdiger Wehner, "Polarized-Light Navigation by Insects," *Scientific American* 235, no. 1 (July 1976): 115.

36. P. Kraft et al., "Honeybee Navigation: Following Routes Using Polarized-Light Cues," *Philosophical Transactions of the Royal Society B* 366, no. 1565 (March 12, 2011): 703–708. https://doi.org/10.1098/rstb.2010.0203.

37. Rüdiger Wehner, "Desert Ant Navigation: How Miniature Brains Solve Complex Tasks," *Journal of Comparative Physiology* 189 (2003): 579–588, https://doi.org/10.1007/s00359-

003-0431-1. Crickets and locusts have analogous functional structures for using polarized light vectors for navigation; however, they appear to detect the polarization through a different method.

38. Karl Von Frisch, *Bees: Their Vision, Chemical Senses and Language* (Ithaca: Cornell University Press), loc. 1021, Kindle.

39. Gould and Gould, *Nature's Compass*, 94.

40. Rachel Muheim, John B. Phillips, and Susanne Akesson, "Polarized Light Cues Underlie Compass Calibration in Migratory Songbirds," *Science* 313 (August 11, 2006): 837–839.

41. Rachel Muheim, Sissel Sjoberg, and Atticus Pinzon-Rodriguez, "Polarized Light Modulates Light-Dependent Magnetic Compass Orientation in Birds," *PNAS* 113, no. 6 (February 9, 2016): 1654, www.pnas.org/cgi/doi/10.1073/pnas.1513391113. Curiously, this same study determined that birds do not use polarized light as an independent compass for orientation, but only for calibrating the magnetic compass.

42. Gould and Gould, *Nature's Compass*, 140–141.

43. Gould and Gould, *Nature's Compass*, 194.

44. Launer, *Navigation Through the Ages*, 3.

45. Dava Sobel, *Longitude* (New York: Penguin Books, 1995), 11–12.

46. Sobel, *Longitude*, 53.

47. Not until the advent of GPS was a worldwide navigation system available that also provides highly accurate position information. Position with GPS is determined based on the time measured for the transmission of signals between orbiting GPS satellites and the user. The timing accuracy of GPS signals is about 25 nanoseconds. This enables GPS to achieve a nominal position accuracy within a few meters.

48. Gould and Gould, *Nature's Compass*, 185–187.

49. Shaun Cain et al., "Magnetic Orientation and Navigation in Marine Turtles, Lobsters and Molluscs: Concepts and Conundrums," *Integrative and Comparative Biology* 45, no. 3 (June 2005): 539.

50. Wiltschko and Wiltschko, "Sensing Magnetic Directions in Birds." This is important because one problem with the magnetic compass is that the information does not enable unique identification of the North and South Pole, but only distinguishes the direction of the pole from the direction of the equator. For birds that migrate across the equator this presents an ambiguity that birds must somehow resolve. Sensing the magnetic inclination angle would enable birds to have continuous navigation when crossing the equator (Wiltschko and Wiltschko, "Magnetic Orientation in Birds," 31).

51. Nathan F. Putman et al., "Longitude Perception and Bicoordinate Magnetic Maps in Sea Turtles," *Current Biology* 21 (March 22, 2011): 463–466, https://doi.org/10.1016/j.cub.2011.01.057.

52. Gould and Gould, *Nature's Compass*, 222.

53. Matthew Collett and Thomas S. Collett, "Insect Navigation: No Map at the End of the Trail?," *Current Biology* 16, no. 2 (January 24, 2006): R48–R51, https://doi.org/10.1016/j.cub.2006.01.007.

54. James F. Cheeseman et al., "Way-Finding in Displaced Clock-Shifted Bees Proves Bees Use a Cognitive Map," *PNAS* 111, no. 24 (June 17, 2014): 8949, www.pnas.org/cgi/doi/10.1073/pnas.1408039111.

55. Randolf Menzel and Uwe Greggers, "The Memory Structure of Navigation in Honeybees," *Journal of Comparative Physiology A* 201 (2015): 559, https://doi.org/10.1007/s00359-015-0987-6.

56. Thomas Alerstam and Anders Hedenstrom, "The Development of Bird Migration Theory," *Journal of Avian Biology* 29, no. 4 (December 1998): 358, https://doi.org/10.2307/3677155.

57. Francisco Pulido, "The Genetics and Evolution of Avian Migration," *Bioscience* 57, no. 2 (February 2007): 167.

58. Gould and Gould, *Nature's Compass*, 10.

59. Hugh Dingle, *Migration: The Biology of Life on the Move* (Oxford: Oxford University Press, 1996), 23–25.

60. Kasper Thorup et al., "Understanding the Migratory Orientation Program of Birds: Extending Laboratory Studies to Study Free-Flying Migrants in a Natural Setting," *Integrative and Comparative Biology* 50, no. 3 (2010): 316. See also Hans G. Wallraff, "Conceptual Approaches to Avian Navigation Systems," in *Orientation in Birds*, ed. Peter Berthold (Basel: Birkhauser Verlag, 1991), 128–165, https://link.springer.com/book/10.1007/978-3-0348-7208-9.

61. Roswitha Wiltschko and Wolfgang Wiltschko, "Avian Navigation: From Historical to Modern Concepts," *Animal Behaviour* 65, no. 2 (February 2003): 257.

62. Gould and Gould, *Nature's Compass*, 136–137.

63. Charles Walcott, "Multi-Model Orientation Cues in Homing Pigeons," *Integrative and Comparative Biology* 45 (2005): 580. The use of olfactory cues has also been confirmed in seabirds. O. Padget et al., "Anosmia Impairs Homing Orientation but Not Foraging Behaviour in Free-Ranging Shearwaters," *Scientific Reports* 7 (August 29, 2017): 9668, https://doi.org/10.1038/s41598-017-09738-5.

64. Gould and Gould, *Nature's Compass*, 121.

65. Hoare, *Animal Migration*, 126–127.

66. Robin Freeman et al., "Predictive Ethoinformatics Reveals the Complex Migratory Behaviour of a Pelagic Seabird, the Manx Shearwater," *Journal of the Royal Society Interface* 10, no. 84 (July 6, 2013): 20130279, http://dx.doi.org/10.1098/rsif.2013.0279.

67. Oliver Padget et al., "Shearwaters Know the Direction and Distance Home but Fail to Encode Intervening Obstacles after Free-Ranging Foraging Trips," *PNAS* 116, no. 43 (October 22, 2019): 21629, https://doi.org/10.1073/pnas.1903829116.

68. Hoare, *Animal Migration*, 141.

69. Brian A. Harrington, "The Hemispheric Globetrotting of the White-Rumped Sandpiper," in *Gatherings of Angels: Migrating Birds and Their Ecology*, ed. Kenneth P. Able (Ithaca: Comstock Books, 1999), 120.

70. Harrington, "The Hemispheric Globetrotting of the White-Rumped Sandpiper," 120.

71. Harrington, "The Hemispheric Globetrotting of the White-Rumped Sandpiper," 129.

72. Harrington, "The Hemispheric Globetrotting of the White-Rumped Sandpiper," 130.

73. Kenneth P. Able, "How Birds Migrate: Flight Behavior, Energetics, and Navigation" in *Gathering of Angels*, 25.

3. NAVIGATIONAL GENIUS—NOT JUST FOR THE BIRDS

1. Bernd Heinrich, *The Homing Instinct: Meaning and Mystery in Animal Migration* (Boston: Houghton Mifflin Harcourt, 2014), 2.

2. Gao Hu et al., "Mass Seasonal Bioflows of High-Flying Insect Migrants," *Science* 354 (December 2016): 1584.

3. Hu et al., "Mass Seasonal Bioflows," 1584.

4. Jurgen Tautz, *The Buzz about Bees: Biology of a Superorganism* (Berlin: Springer-Verlag, 2008), 63.

5. James L. Gould and Carol Grant Gould, *Nature's Compass: The Mystery of Animal Navigation* (Princeton: Princeton University Press, 2012), 77.

6. Stefanie Mares, Lesley Ash, and Wulfila Gronenberg, "Brain Allometry in Bumblebee and Honey Bee Workers," *Brain Behavior and Evolution* 66, no. 1 (2005): 54, https://doi.org/10.1159/000085047.

7. Mares, Ash, and Gronenberg, "Brain Allometry," 59.

8. Ajayrama Kumaraswamy et al., "Adaptations during Maturation in an Identified Honeybee Interneuron Responsive to Waggle Dance Vibration Signals," *eNeuro* 6, no. 5 (2019), https://doi.org/10.1523/ENEURO.0454-18.2019.

9. Veronika Lambinet et al., "Does the Earth's Magnetic Field Serve as a Reference for Alignment of the Honeybee Waggle Dance?," *PLoS ONE* 9, no. 12 (December 26, 2014), https://doi.org/10.1371/journal.pone.0115665.

10. Andrew B. Barron and Jenny Aino Plath, "The Evolution of Honey Bee Dance Communication: A Mechanistic Perspective," *Journal of Experimental Biology* 220, no. 23 (December 2017): 4344, https://doi.org/doi:10.1242/jeb.142778.

11. Barron and Plath, "The Evolution of Honey Bee Dance Communication," 4340.

12. Barron and Plath, "The Evolution of Honey Bee Dance Communication," 4339.

13. Barron and Plath, "The Evolution of Honey Bee Dance Communication." Dwarf honey bee behavior was also described by Karl von Frisch, *Bees: Their Vision, Chemical Senses, and Language* (Ithaca: Cornell University Press, 2014), loc. 1224, Kindle.

14. Martin Lindauer, *Communication Among Social Bees* (Cambridge, MA: Harvard University Press, 1961), cited in John Alcock, *Animal Behavior: An Evolutionary Approach* (Sunderland, MA: Sinauer Associates, 2013), 34.

15. Antoine Wystrach and Paul Graham, "What Can We Learn from Studies of Insect Navigation," *Animal Behaviour* 84, no. 1 (2012): 13–20. In a similar vein, Darwin commented that, "It is certain that there may be extraordinary mental activity with an extremely small absolute mass of nervous matter: thus the wonderfully diversified instincts, mental powers, and affections of ants are notorious, yet their cerebral ganglia are not so large as the quarter of a small pin's head. Under this point of view, the brain of an ant is one of the most marvelous atoms of matter in the world, perhaps more so than the brain of a man." Charles Darwin, *The Descent of Man and Selection in Relation to Sex* (Amherst, NY: Prometheus Books, 1998), 55.

16. Matthew Collett and Thomas S. Collett, "The Learning and Maintenance of Local Vectors in Desert Ant Navigation," *Journal of Experimental Biology* 212, no. 7 (April 2009): 895, https://doi.org/10.1242/jeb.024521.

17. Rüdiger Wehner et al., "Ant Navigation: One-Way Routes Rather than Maps," *Current Biology* 16 (January 10, 2006): 75, https://doi.org/10.1016/j.cub.2005.11.035.

18. Rüdiger Wehner, "The Desert Ant's Navigational Toolkit: Procedural Rather Than Positional Knowledge," *Journal of the Institute of Navigation* 55, no. 2 (Summer 2008), 101.

19. Thomas S. Collett and Matthew Collett, "Route-Segment Odometry and Its Interactions with Global Path-Integration," *Journal of Comparative Physiology A* 201 (2015): 617, https://doi.org/10.1007/s00359-015-1001-z.

20. Stanley Heinze, Ajay Narendra, and Allen Cheung, "Principles of Insect Path Integration," *Current Biology* 28 (September 10, 2018): R1045.

21. Matthias Wittlinger, Rüdiger Wehner, and Harald Wolf, "The Ant Odometer: Stepping on Stilts and Stumps," *Science* 312, no. 5782 (June 2006): 1965, https://doi.org/10.1126/science.1126912.

22. Sarah E. Pfeffer and Matthias Wittlinger, "Optic Flow Odometry Operates Independently of Stride Integration in Carried Ants," *Science* 353 (September 2016): 1155–1157.

23. Collett and Collett, "Route-Segment Odometry," 618.

24. Roman Huber and Markus Knaden, "Egocentric and Geocentric Navigation During Extremely Long Foraging Paths of Desert Ants," *Journal of Comparative Physiology A* 201 (2015): 609, https://doi.org/10.1007/s00359-015-0998-3.

25. Sarah E. Pfeffer and Matthias Wittlinger, "How to Find Home Backwards? Navigation During Rearward Homing of *Cataglyphis fortis* Desert Ants," *Journal of Experimental Biology* 219 (2016): 2119, https://doi.org/10.1242/jeb.137786.

26. Pfeffer and Wittlinger, "How to Find Home Backwards?," 2125.

27. Wehner, "The Desert Ant's Navigational Toolkit," 104.

28. Wehner, commenting on desert ant navigation capabilities, suggests that the experiments lend support to the concept that the ants "use landmarks as contextual cues to recall particular steering commands, i.e., to decide when to do what—or, more generally, that the ant's spatial knowledge is structured in a procedural way rather than in geocentred all-embracing spatial representations." Rüdiger Wehner, "Desert Ant Navigation: How Miniature Brains Solve Complex Tasks," *Journal of Comparative Physiology A* 189 (July 23, 2003): 579–588, https://doi.org/10.1007/s00359-003-0431-1. See also Matthew Collett, Thomas Collett, and Rüdiger Wehner, "Calibration of Vector Navigation in Desert Ants," *Current Biology* 9, no. 18 (September 13, 1999): 1031–1034.

29. Huber and Knaden, "Egocentric and Geocentric Navigation," 609.

30. Mandyam V. Srinivasan, "Where Paths Meet and Cross: Navigation by Path Integration in the Desert Ant and the Honeybee," *Journal of Comparative Physiology A* 201 (2015): 533–546, https://doi.org/10.1007/s00359-015-1000-0.

31. Heinze, Narendra, and Cheung, "Principles of Insect Path Integration," R0143.

32. Richard A. Holland, Martin Wikelski, and David S. Wilcove, "How and Why Insects Migrate," *Science* 313 (August 2006): 794.

33. "Random Sample," *Science* 343, no. 6171 (February 7, 2014): 584, https://doi.org/10.1126/science.343.6171.583-c. As noted, the observed population of monarchs in Mexico has decreased significantly in recent years. It appears this is likely caused by a decrease in the availability of milkweed.

34. Michelle J. Solensky, "Overview of Monarch Migration," in *The Monarch Butterfly: Biology and Conservation*, eds. Karen S. Oberhauser and Michelle J. Solensky (Ithaca, NY: Cornell University Press, 2004), 81.

35. Sandra M. Perez and Orley R. Taylor, "A Sun Compass in Monarch Butterflies," *Nature* 387 (May 1997): 29.

36. Christine Merlin, Robert J. Gegear, and Steven M. Reppert, "Antennal Circadian Clocks Coordinate Sun Compass Orientation in Migratory Monarch Butterflies," *Science* 325 (September 2009): 1700.

37. A good description of the mechanisms involved in monarch navigation is contained in the paper by Steven M. Reppert, Patrick A. Guerra, and Christine Merlin, "Neurobiology of Monarch Butterfly Migration," *Annual Review of Entomology* 61 (2016): 25–42.

38. Shuai Zhan et al., "The Monarch Butterfly Genome Yields Insights into Long-Distance Migration," *Cell* 147 (November 2011): 1171–1185; Shuai Zhan et al., "The Genetics of Monarch Butterfly Migration and Warning Colouration," *Nature* 314 (October 2014): 317–321, https://doi.org/10.1038/nature13812.

39. Reppert et al., "Neurobiology of Monarch Butterfly Migration," 37.

40. Eli Shlizerman et al., "Neural Integration Underlying a Time-Compensated Sun Compass in the Migratory Monarch Butterfly," *Cell Reports* 15, no. 4 (April 26, 2016): 683–691, https://doi.org/10.1016/j.celrep.2016.03.057.

41. Reppert et al., "Neurobiology of Monarch Butterfly Migration," 16.

42. Marie Dacke et al., "Dung Beetles Use the Milky Way for Orientation," *Current Biology* 23, no. 4 (2013): 298, https://doi.org/10.1016/j.cub.2012.12.034.

43. Basil el Jundi et al., "Neural Coding Underlying the Cue Preference for Celestial Orientation," *PNAS* 112, no. 36 (September 2015): 11395, https://doi.org/10.1073/pnas.1501272112.

44. Marie Dacke, Peter Nordstrom, and Clarke H. Scholtz, "Twilight Orientation to Polarized Light in the Crepuscular Dung Beetle *Scarabaeus zambesianus*," *Journal of Experimental Biology* 206 (2003): 1535, https://doi.org/10.1242/jeb.00289. The special eye mechanism is called a rhabdom in the dorsal rim area (DRA) of the eye.

45. el Jundi et al., "Neural Coding," 11395.

46. The name is derived from a type of seaweed (*Sargassum*) that is plentiful in this part of the ocean.

47. Gould and Gould, *Nature's Compass*, 217.

48. Nathan F. Putman et al., "Longitude Perception and Bicoordinate Magnetic Maps in Sea Turtles," *Current Biology* 21, no. 6 (2011): 463, https://doi.org/10.1016/j.cub.2011.01.057.

49. Putman et al., "Longitude Perception," 463.

50. Miriam Liedvogel, Susanne Akeeson, and Staffan Bensch, "The Genetics of Migration on the Move," *Trends in Ecology and Evolution* 26, no. 11 (November 2011): 565.

51. Gould and Gould, *Nature's Compass*, 230.

52. Hugh Dingle, *Migration: The Biology of Life on the Move* (Oxford: Oxford University Press, 1996), 304–305.

53. Peter Berthold, "A Comprehensive Theory for the Evolution, Control and Adaptability of Avian Migration," *Ostrich* 70 (1999): 1–11, cited in Francisco Pulido, "The Genetics and Evolution of Avian Migration," *Bioscience* 57, no. 2 (February 2007): 167.

54. Derek A. Roff and Daphne J. Fairbairn, "The Evolution and Genetics of Migration in Insects," *BioScience* 57, no. 2 (February 2007): 155–164.

55. Dingle, *Migration: The Biology of Life on the Move*, 305.

56. Dingle, *Migration: The Biology of Life on the Move*, 320.

57. Pulido, "The Genetics and Evolution of Avian Migration," 169.

58. Pulido, "The Genetics and Evolution of Avian Migration," 166.

59. William J. Sutherland, "Evidence for Flexibility and Constraint in Migration Systems," *Journal of Avian Biology* 29 (1998): 441.

60. Thomas Alerstam and Anders Hedenstrom, "The Development of Bird Migration Theory," *Journal of Avian Biology* 29 (1998): 360.

61. Pulido, "The Genetics and Evolution of Avian Migration," 170.

62. Pulido, "The Genetics and Evolution of Avian Migration," 167.

63. Pulido, "The Genetics and Evolution of Avian Migration," 170.

64. Günter P. Wagner and Vincent J. Lynch, "Evolutionary Novelties," *Current Biology* 20, no. 2 (2010): R48.

65. Gerd B. Müller, "Epigenetic Innovation," in *Evolution: The Extended Synthesis*, eds. Massimo Pigliucci and Gerd Muller (Cambridge, MA: MIT Press, 2010), 310.

66. Douglas Futuyma, *Evolution*, 3rd ed. (Sunderland, MA: Sinauer Associates, Inc., 2013), 200.

67. Mark Kirkpatrick, "Genes and Adaptation: A Pocket Guide to the Theory," in *Adaptation*, eds. Michael R. Rose and George V. Lauder (San Diego: Academic Press, 1996), 141–142.

68. Stephen Jay Gould and Richard Lewontin, "The Spandrels of San Marco and the Panglossian Paradigm: A Critique of the Adaptationist Programme," *Proceedings of the Royal Society B* 205, no. 1161 (September 21, 1979), https://doi.org/10.1098/rspb.1979.0086.

69. Jerry Coyne, "The Great Mutator," review of *The Edge of Evolution*, by Michael J. Behe, *The New Republic* (June 18, 2007): 39.

70. Jerry Fodor and Massimo Piattelli-Palmarini, *What Darwin Got Wrong* (New York: Farrar, Straus and Giroux, 2010), 21.

71. Tim Lewens, *Organisms and Artifacts: Design in Nature and Elsewhere* (Cambridge, MA: MIT Press, 2004), 33.

72. Pulido, "The Genetics and Evolution of Avian Migration," 171. Pulido writes, "Not all traits are equally likely to covary genetically. Behavioral and physiological traits that are directly linked to the migratory journey—including the amount and timing of migratory activity, the timing of molt, the endogenously preferred direction of movement, and the amount of energy stored before the migratory journey—seem to be strongly correlated. Other characters of the migratory syndrome, such as morphological and life-history traits, do not tend to covary genetically with migratory behavior, or do so only loosely."

73. Gould and Gould, *Nature's Compass*, 185–187.

74. Implicit memory involves information that can be recalled automatically. An example is procedural knowledge, such as human motor skills. In animals, implicit procedural memory involves information used in complex programmed behaviors such as navigation, nest building, and communication.

75. Christine Merlin and Miriam Liedvogel, "The Genetics and Epigenetics of Animal Migration and Orientation: Birds, Butterflies and Beyond," *Journal of Experimental Biology* 222, suppl. 1 (February 6, 2019): 2, https://doi.org/10.1242/jeb.191890.

76. Merlin and Liedvogel, "The Genetics and Epigenetics of Animal Migration," 1.

77. William W. Cochran, Henrik Mouritsen, and Martin Wikelski, "Migrating Songbirds Recalibrate their Magnetic Compass Daily from Twilight Cues," *Science* 304 (April 2004): 405.

4. COMPLEX PROGRAMMED SOCIETIES

1. J. Henri Fabre, *The Insect World of J. Henri Fabre*, trans. Alexander Teixeira de Mattos (Boston: Beacon Press, 1991), 65.

2. Peter Skorupski, Johannes Spaethe, and Lars Chittka, "Visual Search and Decision Making in Bees: Time, Speed, and Accuracy," *International Journal of Comparative Psychology* 19 (2006) 19: 342.

3. David Westneat and Charles Fox, eds., *Evolutionary Behavioral Ecology* (Oxford: Oxford University Press, 2010), 286–287.

4. Karen M. Kapheim, "Tinbergen's Four Questions and the Future of Sociogenomics," *Behavioral Ecology and Sociobiology* 72 (2018): 186, https://doi.org/10.1007/s00265-018-2606-3.

5. Kapheim, "Tinbergen's Four Questions," 186.

6. Wilson defines sociobiology as the "systematic study of the biological basis of all social behavior." Edward O. Wilson, *Sociobiology: The New Synthesis* (Cambridge, MA: Belknap Press, 2000), glossary.

7. Bert Hölldobler and E. O. Wilson, *The Superorganism: The Beauty, Elegance, and Strangeness of Insect Societies* (New York: W. W. Norton and Co., 2009), 4.

8. "Social Insects," in *Grzimek's Animal Life Encyclopedia*, 2nd ed., vol. 3, *Insects*, ed. Neil Schlager (Detroit: Thomson Gale, 2004), 68.

9. Wyatt A. Shell and Sandra M. Rehan, "Behavioral and Genetic Mechanisms of Social Evolution: Insights from Incipiently and Facultatively Social Bees," *Apidologie* 49 (2018): 13–30, https://doi.org/10.1007/s13592-017-0527-1.

10. Direct fitness is "a measure of the reproductive or genetic success of an individual based on the number of offspring that live to reproduce." This contrasts with "indirect fitness," which is "a measure of the genetic success of an altruistic individual based on the number of relatives (or genetically similar individuals) that the altruist helps reproduce that would not otherwise have survived to do so." Indirect fitness in theory applies to workers. John Alcock, *Animal Behavior: An Evolutionary Approach* (Sunderland, MA: Sinauer Associates, 2013), 23.

11. Hölldobler and Wilson, *The Superorganism*, 136–137.

12. Amy L. Toth and Sandra M. Rehan, "Molecular Evolution of Insect Sociality: An Eco-Evo-Devo Perspective," *Annual Review of Entomology* 62 (2017): 424.

13. Westneat and Fox, *Evolutionary Behavioral Ecology*, 328.

14. Brian R. Johnson and Timothy A. Linksvayer, "Deconstructing the Superorganism: Social Physiology, Groundplans, and Sociogenomics," *The Quarterly Review of Biology* 85, no. 1 (March 2010): 73.

15. Gene E. Robinson, Russell D. Fernald, and David F. Clayton, "Genes and Social Behavior," *Science* 322, no. 5903 (November 2008): 896.

16. Thomas D. Seeley, "The Honey Bee Colony as a Superorganism," in *Exploring Animal Behavior: Readings from American Scientist*, eds. Paul W. Sherman and John Alcock (Sunderland, MA: Sinauer Associates, 2010), 369.

17. Seeley, "The Honey Bee Colony as a Superorganism," 370.

18. Hölldobler and Wilson, *The Superorganism*, 117. Hölldobler and Wilson write, "Watched for only a few hours, a colony of social insects might be interpreted as consisting of automata driven with the same uniform set of decision rules. But that is far from the case. Each member of the colony is distinct in some manner or other that affects its behavior. Each has a mind of its own. By mind we do not mean a reflective, self-aware, wide-roaming consciousness of the human mind, but rather a cognitive consciousness built with a relatively complex brain that can store information from all its sensory modalities (taste, smell, touch, sight, and sound) as well as some memory of the events it has experienced in its short life. Labor is driven by priorities influenced at least in part by prior cognition: search for this task, or finish the task in progress, or patrol the surround in search for whatever needs doing, or pause to stand guard, or simply rest."

19. Elva J. H. Robinson, Ofer Feinerman, and Nigel R. Franks, "Flexible Task Allocation and the Organization of Work in Ants," *Proceedings of the Royal Society B* 276 (2009): 4373.

20. Deborah Gordon, "The Organization of Work in Social Insect Colonies," *Nature* 380 (March 1996): 121.

21. Hölldobler and Wilson, *The Superorganism*, 115.

22. Samuel N. Beshers and Jennifer H. Fewell, "Models of Division of Labor in Insects," *Annual Review of Entomology* 46, no. 1 (2001): 413–440.

23. Robinson, Feinerman, and Franks, "Flexible Task Allocation," 4373.

24. Hölldobler and Wilson, *The Superorganism*, 93.

25. Joan E. Strassmann and David C. Queller, "Insect Societies as Divided Organisms: The Complexities of Purpose and Cross-Purpose," in *In the Light of Evolution*, vol. 1, *Adaptation and Complex Design*, eds. John C. Avise and Francisco J. Ayala (Washington: National Academies Press, 2007), 163.

26. Joshua P. Martin et al., "The Neurobiology of Insect Olfaction: Sensory Processing in a Comparative Context," *Progress in Neurobiology* (2011): 427.

27. Hölldobler and Wilson, *The Superorganism*, 180.

28. Tristram D. Wyatt, *Pheromones and Animal Behavior: Chemical Signals and Signatures* (Cambridge, MA: Cambridge University Press, 2014),195.

29. Wyatt, *Pheromones and Animal Behavior*, 15.

30. Wyatt, *Pheromones and Animal Behavior*, 18.

31. Mary Jane West-Eberhard, *Developmental Plasticity and Evolution* (Oxford: Oxford University Press, 2003), 246.

32. Hölldobler and Wilson, *The Superorganism*, 179.

33. Bert Hölldobler, "The Chemistry of Social Regulation: Multicomponent in Ant Societies," *PNAS* 92 (January 1995): 19.

34. Hölldobler, "The Chemistry of Social Regulation," 21.

35. Wyatt, *Pheromones and Animal Behavior*, 28.

36. One formal definition of altruism is "helpful behavior that raises the recipient's direct fitness while lowering the donor's direct fitness." Alcock, *Animal Behavior: An Evolutionary Approach*, glossary.

37. Hölldobler and Wilson, *The Superorganism*, 6.

38. Edward O. Wilson, *The Social Conquest of Earth* (New York: Liveright Publishing, 2012), 110.

39. E. J. Fittkau and H. Klinge, "On Biomass and Trophic Structure of the Central Amazonian Rain Forest Ecosystem," *Biotropica* 5, no. 1 (1973): 2–14, https://doi.org/10.2307/2989676.

40. Wilson, *The Social Conquest of Earth*, 109.

41. Strassmann and Queller, "Insect Societies as Divided Organisms," 153.

42. S. Hollis Woodard et al., "Genes Involved in Convergent Evolution of Eusociality in Bees," *PNAS* 108, no. 18 (May 3, 2011): 7472.

43. Jürgen Tautz, *The Buzz about Bees: Biology of a Superorganism*, trans. David C. Sandeman (Berlin: Springer-Verlag, 2008), 3.

44. Tautz, *The Buzz About Bees*, 46.

45. Guy Bloch and Christina Grozinger, "Social Molecular Pathways and the Evolution of Bee Societies," *Philosophical Transactions of the Royal Society B* 366 (2011): 2157, https://doi.org/10.1098/rstb.2010.0346.

46. Hölldobler and Wilson, *The Superorganism*, 100.

47. Gould and Gould, *Animal Architects*, 104.

48. Thomas D. Seeley, *Honeybee Democracy* (Princeton, NJ: Princeton University Press, 2010), 66.

49. Thomas D. Seeley, "Honey Bee Colonies are Group-Level Adaptive Units," *American Naturalist* 150 Supplement (July 1997): S22.

50. Seeley, *Honeybee Democracy*, 27.

51. Seeley, *Honeybee Democracy*, 27.

52. Jane Memmott, Nickolas M. Waser, and Mary V. Price, "Tolerance of Pollination Networks to Species Extinctions," *Proceedings of the Royal Society of London* 271 (2004): 2605.

53. Tautz, *The Buzz About Bees*, 57. Of the flowering plants pollinated by insects, about 85 percent are pollinated by honey bees (approximately 170,000 species of flowering plants), including 90 percent of fruit trees. The implication is that "this extreme numerical imbalance between plant clients and pollinators is remarkable, and suggests that honey bees are so successful in this area as to leave little room for the coexistence of similarly occupied competitors."

54. Tautz, *The Buzz About Bees*, 3.

55. Hölldobler and Wilson, *The Superorganism*, 4.

56. Bert Hölldobler and E. O. Wilson, *The Leafcutter Ants* (New York: W. W. Norton, 2011), 3.

57. Hölldobler and Wilson, *The Superorganism*, 97.

58. Nicholas Davies, John Krebs, and Stuart West, *An Introduction to Behavioural Ecology* (Oxford: Wiley-Blackwell, 2012), 362.

59. Hölldobler and Wilson, *The Superorganism*, 145.

60. Hölldobler and Wilson, *The Superorganism*, 100.

61. Hölldobler and Wilson, *Journey to the Ants* (Cambridge, MA: Belknap Press, 1994), 40.

62. Hölldobler and Wilson, *Journey to the Ants*, 29.

63. Hölldobler and Wilson, *Journey to the Ants*, 29.

64. Hölldobler and Wilson, *Journey to the Ants*, 30.

65. Hölldobler and Wilson, *The Leafcutter Ants*, 1.

66. Hölldobler and Wilson, *The Superorganism*, 163.

67. Hölldobler and Wilson, *The Superorganism*, 163.

68. J. Roschard and F. Roces, "Cutters, Carriers and Transport Chains: Distance-Dependent Foraging Strategies in the Grass-Cutting Ant *Atta vollenweederi*," *Insectes Sociaux* 50, no. 3 (2003): 237–244, https://doi.org/10.1007/s00040-003-0663-7.

69. Ulrich Mueller, interviewed by Leah Ariniello, "Protecting Paradise: Fungus-Farming Ants Ensure Crop Survival with Surprising Strategies and Partnerships," *Bioscience* 49, no. 10 (October 1999): 763, https://doi.org/10.2307/1313566.

70. Hölldobler and Wilson, *The Leafcutter Ants*, 32.

71. Hölldobler and Wilson, *The Leafcutter Ants*, 90.

72. Michael Poulsen and Jacobus J. Boomsma, "Mutualistic Fungi Control Crop Diversity in Fungus-Growing Ants," *Science* 307, no. 5710 (February 2005): 743.

73. Poulsen and Boomsma, "Mututalistic Fungi," 742.

74. Hölldobler and Wilson, *The Leafcutter Ants*, 95.

75. Hölldobler and Wilson, *The Leafcutter Ants*, 94.

76. Michael L. Cain, William D. Bowman, and Sally D. Hacker, *Ecology* (Sunderland, MA: Sinauer Associates, 2011), 320.

77. "Isoptera (Termites)," *Grzimek's Animal Life Encyclopedia*, 165.

78. "Isoptera (Termites)," *Grzimek's Animal Life Encyclopedia*, 168.

79. "Isoptera (Termites)," *Grzimek's Animal Life Encyclopedia*, 166.

80. George McGhee, *Convergent Evolution: Limited Forms Most Beautiful* (Cambridge, MA: MIT Press, 2011), 224.

81. Michael Poulsen et al., "Complementary Symbiont Contributions to Plant Decomposition in a Fungus-Farming Termite," *PNAS* 111, no. 40 (October 7, 2014): 14500.

82. Poulsen et al., "Complementary Symbiont Contributions," 14501.

83. West-Eberhard, *Developmental Plasticity and Evolution*, 34. West-Eberhard's definition for phenotypic plasticity is, "The ability of an organism to react to an environmental input with a change in form, state, movement, or rate of activity."

84. Charles Darwin, *On the Origin of Species: By Means of Natural Selection, or the Preservation of Favoured Races in the Struggle for Life* (London: John Murray, 1859), 184. Darwin writes, "As we sometimes see individuals of a species following habits widely different from those both of their own species and of the other species of the same genus, we might expect, on my theory, that such individuals would occasionally have given rise to new species, having anomalous habits, and with their structure either slightly or considerably modified from that of their proper type."

85. Emilie C. Snell-Rood, "An Overview of the Evolutionary Causes and Consequences of Behavioural Plasticity," *Animal Behaviour* 85 (May 2013): 1005.

86. Snell-Rood, "An Overview," 1004.

87. Frederic Mery and James G. Burns, "Behavioural Plasticity: An Interaction Between Evolution and Experience," *Evolutionary Ecology* 24 (2010): 574, https://doi.org/10.1007/s10682-009-9336-y.

88. Snell-Rood, "An Overview," 1006.

89. Reuven Dukas, "Evolutionary Biology of Insect Learning," *Annual Review of Entomology* 53 (2008): 155.

90. Stefanie Mares, Lesley Ash, and Wulfila Gronenberg, "Brain Allometry in Bumblebee and Honey Bee Workers," *Brain, Behavior and Evolution* 66, no. 1 (2005): 54, https://doi.org/10.1159/000085047.

91. Lawrence D. Harder and Leslie A. Real, "Why are Bumble Bees Risk Averse?" *Ecology* 68, no. 4 (August 1987): 1104.

92. Tautz, *The Buzz about Bees*, 72–73.

93. Tautz, *The Buzz about Bees*, 84.

94. David F. Sherry and John B. Mitchell, "Neuroethology of Foraging," in *Foraging: Behavior and Ecology*, eds. David Stephens et al. (University of Chicago Press, Chicago, 2007), 64.

95. Jonathan Cnaani, James D. Thomson, and Daniel R. Papaj, "Flower Choice and Learning in Foraging Bumblebees: Variation in Nectar Volume and Concentration," *Ethology* 112 (2006): 278–285.

96. J. L. Gould, "Natural History of Honey Bee Learning," in *The Biology of Learning*, ed. Peter Marler and H. S. Terrace (Berlin: Springer-Verlag, 1984), 161.

97. Cwyn Solvi, Selene Gutierrez Al-Khudhairy, and Lars Chittka, "Bumble Bees Display Cross-Modal Object Recognition Between Visual and Tactile Senses," *Science* 367 (February 2020): 910.

98. Makoto Mizunami, Yoshitaka Hamanaka, and Hiroshi Nishino, "Toward Elucidating Diversity of Neural Mechanisms Underlying Insect Learning," *Zoological Letters* 1 (2015): 8, https://doi.org/ 10.1186/s40851-014-0008-6.

99. Mery and Burns, "Behavioural Plasticity."

100. Mery and Burns, "Behavioural Plasticity."

101. Peter Godfrey-Smith, "Baldwin Skepticism and Baldwin Boosterism," in *Evolution and Learning: The Baldwin Effect Reconsidered*, eds. Bruce H. Weber and David J. Depew (Cambridge, MA: MIT Press, 2007), 54–55.

102. Stefano Ghirlanda, Magnus Enquist, and Johan Lind, "Coevolution of Intelligence, Behavioral Repertoire, and Lifespan," *Theoretical Population Biology* 91 (2014): 44–49.

103. Beverly J. Piggott et al., "The Neural Circuits and Synaptic Mechanisms Underlying Motor Initiation in C. *elegans*," *Cell* 147, no. 4 (November 11, 2011): 922.

104. Catharine H. Rankin, "Invertebrate Learning: What Can't a Worm Learn?" *Current Biology* 14, no. 15 (August 10, 2004): R617-R618.

105. Rankin, "Invertebrate Learning."

106. Piggott et al., "The Neural Circuits," 922.

107. William Schafer, "Nematode Nervous Systems," *Current Biology* 26 (October 24, 2016): R957.

108. Hölldobler and Wilson, *The Superorganism*, 54.

109. Hölldobler and Wilson, *The Superorganism*, 55.

110. Hölldobler and Wilson, *The Superorganism*, 6.

111. Hölldobler and Wilson, *The Superorganism*, 7.

112. Hölldobler and Wilson, *The Superorganism*, 57.

113. Hölldobler and Wilson argue that, "As rules are multiplied, the potential outcomes increase at a superexponential rate. In an insect colony, one algorithm with three successive

decision points has only eight outcomes. But seven such algorithms in combination have over 2 million outcomes. From such a vast pool of potential outcomes, natural selection has chosen a microscopically small set of algorithms and algorithmic combinations." Hölldobler and Wilson, *The Superorganism*, 58.

114. Joan E. Strassmann and David C. Queller, "Insect Societies as Divided Organisms," 152.

115. Hua Yan et al., "Eusocial Insects as Emerging Models for Behavioural Epigenetics," *Nature Reviews: Genetics* 15 (October 2014): 678.

116. Brian R. Johnson et al., "Phylogenomics Resolves Evolutionary Relationships among Ants, Bees, and Wasps," *Current Biology* 23, no. 20 (October 21, 2013): 2059, https://doi.org/10.1016/j.cub.2013.08.050.

117. Toth and Rehan, "Molecular Evolution of Insect Sociality," 420.

118. Bloch and Grozinger, "Social Molecular Pathways," 2155.

119. Hölldobler and Wilson, *The Superorganism*, 39.

120. Bloch and Grozinger, "Social Molecular Pathways," 2162.

121. Johnson and Linksvayer, "Deconstructing the Superorganism," 66.

122. Johnson and Linksvayer, "Deconstructing the Superorganism," 66.

123. Johnson and Linksvayer, "Deconstructing the Superorganism," 68.

124. The formal definition of orphan gene is "a previously unidentified protein-coding open reading frame, that has no clear-cut homolog in any organisms," *Dictionary of Genetics*, 7th ed. (Oxford: Oxford University Press, 2006), s.v. "orphan gene."

125. Konstantin Khalturin et al., "More Than Just Orphans: Are Taxonomically-Restricted Genes Important in Evolution?," *Trends in Genetics* 25, no. 9 (2009): 404.

126. Diethard Tautz and Tomislav Domazet-Loso, "The Evolutionary Origin of Orphan Genes," *Nature Reviews: Genetics* 12 (October 2011): 700.

127. Tautz and Domazet-Loso, "The Evolutionary Origin of Orphan Genes," 700.

128. Khalturin et al., "More Than Just Orphans," 406.

129. Henrik Kaessmann, "Origins, Evolution, and Phenotypic Impact of New Genes," *Genome Research* 20 (July 22, 2010): 1313.

130. Hölldobler and Wilson, *The Superorganism*, 41.

131. Johnson and Linksvayer, "Deconstructing the Superorganism," 73.

132. Woodard et al., "Genes Involved in Convergent Evolution," 7473.

133. Woodard et al., "Genes Involved in Convergent Evolution," 7473.

134. Karen M. Kapheim et al., "Genomic Signatures of Evolutionary Transitions from Solitary to Group Living," *Science* 348 (June 2015): 1139–1143.

135. Kapheim et al., "Genomic Signatures," 1139.

136. Julia C. Jones et al., "Extreme Differences in Recombination Rate between the Genomes of a Solitary and a Social Bee," *Molecular Biology and Evolution* 36, no. 10 (2019): 2277–2291, https://doi.org/10.1093/molbev/msz130.

137. Doori Park et al., "Uncovering the Novel Characteristics of Asian Honey Bee, *Apis cerana*, by Whole Genome Sequencing," *BMC Genomics* 16, no. 1 (2015): 5, https://doi.org/10.1186/1471-2164-16-1.

138. Park et al., "Uncovering," 1.

139. Park et al., "Uncovering," 2.

140. Pedro G. Ferreira et al., "Transcriptome Analyses of Primitively Eusocial Wasps Reveal Novel Insights into the Evolution of Sociality and the Origin of Alternative Phenotypes," *Genome Biology* 14 (2013): R20.

141. B. Feldmeyer, D. Elsner, and S. Foitzik, "Gene Expression Patterns Associated with Caste and Reproductive Status in Ants: Worker-Specific Genes are More Derived than Queen-Specific Ones," *Molecular Ecology* 23 (2014): 151–161, https://doi.org/10.1111/mec.12490.

142. Seirian Sumner, "The Importance of Genomic Novelty in Social Evolution," *Molecular Ecology* 23 (2014): 27.

143. Daniel F. Simola et al., "Social Insect Genomes Exhibit Dramatic Evolution in Gene Composition and Regulation While Preserving Regulatory Features Linked to Sociality," *Genome Research* 23 (2013): 1235.

144. Simola et al., "Social Insect Genomes," 1237.

145. Non-coding refers to regions of the genome that do not code for proteins.

146. Benjamin E. R. Rubin et al., "Rate Variation in the Evolution of Non-Coding DNA Associated with Social Evolution in Bees," *Philosophical Transactions of the Royal Society B* 374, no. 1777 (2019): 20180247, https://doi.org/10.1098/rstb.2018.0247.

147. Rubin et al., "Rate Variation," 9.

148. Craig B. Lowe et al., "Three Periods of Regulatory Innovation During Vertebrate Evolution," *Science* 333, no. 6045 (August 2011): 1019–1024.

149. Conrad H. Waddington, "The Epigenotype," *Endeavour* 1 (1942): 18–20.

150. Cris C. Ledón-Rettig, Christina L. Richards, and Lynn B. Martin, "Epigenetics for Behavioral Ecologists," *Behavioral Ecology* 24, no. 2 (March-April 2013), under "Introduction," https://doi.org/10.1093/beheco/ars145. Emphasis in original.

151. Jonathan Wells, "Membrane Patterns Carry Ontogenetic Information That Is Specified Independently of DNA," *BIO-Complexity* 2 (2014): 15, https://doi.org/10.5048/BIO-C.2014.

152. Wells, "Membrane Patterns," 16.

153. T. Bas Rodenburg, "The Role of Genes, Epigenetics and Ontogeny in Behavioural Development," *Applied Animal Behaviour Science* 157 (2014): 8, https://doi.org/10.1016/j.applanim.2014.06.002.

154. Hua Yan et al., "Eusocial Insects as Emerging Models for Behavioural Epigenetics," *Nature Reviews: Genetics* 15 (October 2014): 681.

155. Yan et al., "Eusocial Insects," 681.

156. Darwin, *On the Origin of Species*, 236–237.

157. Richard Dawkins, *The Selfish Gene* (Oxford: Oxford University Press, 1989).

158. W. D. Hamilton, "The Genetical Evolution of Social Behaviour," *Journal of Theoretical Biology* 7 (1964): 1–16.

5. Insect Architecture

1. E. B. White, *Charlotte's Web* (New York: HarperTrophy, 2012), loc. 108, Kindle.

2. Bernd Heinrich, *The Homing Instinct: Meaning and Mystery in Animal Migration* (Boston: Houghton Mifflin Harcourt, 2014), 125.

3. James L. Gould and Carol Grant Gould, *Animal Architects* (New York: Basic Books, 2007), 4.

4. Andrea Perna and Guy Theraulaz, "When Social Behaviour is Moulded in Clay: On Growth and Form of Social Insect Nests," *Journal of Experimental Biology* 220 (2017): 83.

5. Charles Darwin, *On the Origin of Species: By Means of Natural Selection, or the Preservation of Favoured Races in the Struggle for Life* (London: John Murray, 1859), 224.

6. Sophie Cardinal and Bryan N. Danforth, "The Antiquity and Evolutionary History of Behavior in Bees," *PLoS ONE* 6, no. 6 (June 2011): e21086.

7. Gould and Gould, *Animal Architects*, 79.

8. Gould and Gould, *Animal Architects*, 100.

9. Heinrich, *The Homing Instinct*, 111.

10. Gould and Gould, *Animal Architects*, 108.

11. Gould and Gould, *Animal Architects*, 108.

12. Gould and Gould, *Animal Architects*, 100. In addition, the Goulds write, "Many species also make wax in special glands unique to bees; this substance may be added to batumen to achieve even better results, particularly at higher temperatures. Beeswax does not sag until well over 100°F, and it is kept carefully below this level by evaporative cooling. Other species use the wax as a thin layer of waterproof varnish around tunnels and cells."

13. Heinrich, *The Homing Instinct*, 127.

14. Jürgen Tautz, *The Buzz about Bees: Biology of a Superorganism*, trans. David C. Sandeman (Berlin: Springer-Verlag, 2008), 159, 160.

15. Tautz, *The Buzz about Bees*, 161.

16. Heinrich, *The Homing Instinct*, 128.

17. Karl von Frisch, *Animal Architecture* (New York: Harcourt Brace Jovanovich, 1974), 93.

18. Nuru Adgaba et al., "An Experiment on Comb Orientation by Honey Bees (*Hymenoptera: Apidae*) in Traditional Hives," *Journal of Economic Entomology* 105, no. 3 (June 2021): 777–782, https://doi.org/10.1603/EC11410.

19. Heinrich, *The Homing Instinct*, 128; Adriane Alexandre Machado De-Melo et al., "Composition and Properties of Apis Mellifera Honey: A Review," *Journal of Apicultural Research* (June 2017), https://doi.org/10.1080/00218839.2017.1338444; Elise Mandl, "Does Honey Ever Go Bad? What You Should Know," Healthline, May 20, 2018, https://www.healthline.com/nutrition/does-honey-go-bad.

20. Gould and Gould, *Animal Architects*, 101.

21. Darwin, *On the Origin of Species*, 235.

22. Gould and Gould, *Animal Architects*, 17.

23. Gould and Gould, *Animal Architects*, 17.

24. Gould and Gould, *Animal Architects*, 87.

25. Douglas H. Chadwick, "Sisterhood of Weavers," *National Geographic* (May 2011): 89.

26. Gould and Gould, *Animal Architects*, 97.

27. See, for example, the excellent weaver ant video at the Aus Ants website, https://ausants.net/documentaries; or this Smithsonian video, https://www.youtube.com/watch?v=iPxxW2UhKEY.

28. Bert Hölldobler and E. O. Wilson, "The Evolution of Communal Nest-Weaving in Ants," in *Exploring Animal Behavior*, eds. Paul W. Sherman and John Alcock (Sunderland, MA: Sinauer Associates, 2010), 138.

29. Hölldobler and Wilson, "The Evolution of Communal Nest-Weaving in Ants," 146.

30. Hölldobler and Wilson, "The Evolution of Communal Nest-Weaving in Ants," 146.

31. Hölldobler and Wilson, "The Evolution of Communal Nest-Weaving in Ants," 147.

32. Gould and Gould, *Animal Architects*, 136.

33. Bert Hölldobler and E. O. Wilson, *The Superorganism: The Beauty, Elegance, and Strangeness of Insect Societies* (New York: W. W. Norton. 2009), 60.

34. Samuel A. Ocko et al., "Solar-Powered Ventilation of African Termite Mounds," *Journal of Experimental Biology* 220, no. 18 (September 2017): 3263, https://doi.org/10.1242/jeb.160895.

35. Kamaljit Singh et al., "The Architectural Design of Smart Ventilation and Drainage Systems in Termite Nests," *Science Advances* 5, no. 3 (March 2019): 1.

36. Samuel A. Ocko, Alexander Heyde, and L. Mahadevan, "Morphogenesis of Termite Mounds," *PNAS* 116, no. 9 (February 2019): 3379, https://doi.org/10.1073/pnas.1818759116.

37. von Frisch, *Animal Architecture*, 138.

38. Juan A. Bonachela et al., "Termite Mounds Can Increase the Robustness of Dryland Ecosystems to Climate Change," *Science* 347, no. 6222 (February 2015): 651.

39. Elizabeth Pennisi, "Africa's Soil Engineers: Termites," *Science* 347 (February 2015): 597.

40. Gould and Gould, *Animal Architects*, 144.

41. J. Scott Turner, *Purpose and Desire: What Makes Something "Alive" and Why Modern Darwinism Has Failed to Explain It* (New York: HarperOne, 2017), 5.

42. Elizabeth Pennisi, "Untangling Spider Biology," *Science* 358 (October 2017): 292.

43. Barbara Taylor, Jen Green, and John Farndon, *The Big Bug Book* (London: Hermes House, 2007), 22.

44. Gould and Gould, *Animal Architects*, 19.

45. Robert Service notes that "mass-produced, superstrong fibers remain out of reach." Robert F. Service, "Silken Promises," *Science* 358, no. 6361 (October 2017): 293–294.

46. Taylor, Green, and Farndon, *The Big Bug Book*, 24.

47. Heinrich, *The Homing Instinct*, 183.

48. Valerie Altounian, Elizabeth Pennisi, and Robert F. Service, "A Spinner's Secrets," *Science* 358 (October 2017): 292, https://doi.org/10.1126/science.358.6361.292.

49. Two such videos are Deep Look's "Is a Spider's Web a Part of Its Mind?," https://www.youtube.com/watch?v=rpwkgMX4IlQ; and BBC Earth's "Beautiful Spider Web Build Time-Lapse," https://www.youtube.com/watch?v=zNtSAQHNONo.

50. Gould and Gould, *Animal Architects*, 52–53.

51. William Shear, "Untangling the Evolution of the Web," in *Exploring Animal Behavior*, eds. Sherman and Alcock, 149.

52. Gould and Gould, *Animal Architects*, 50. The Goulds further explain, "The vibrations induced by struggling insects are transmitted along the slack and sticky spiral of catching threads to the taut radii, and thence to the center. The spider compares the strength of vibrations between radii to interpolate the angle of the prey from the center. In theory, the

spider should simply rush out along the closest radius until it encounters the insect. However, when vibrations are induced experimentally, the spider nevertheless stops at about the right distance out from the center of the web."

53. Heinrich, *The Homing Instinct*, 50.

54. Shear, "Untangling the Evolution of the Web," 148–158.

55. Shear, "Untangling the Evolution of the Web," 151.

56. Shear, "Untangling the Evolution of the Web," 153–154.

57. Pennisi, "Untangling Spider Biology," 288–291.

58. Richard Dawkins, *The Blind Watchmaker*, 2nd ed. (New York: W.W. Norton, 1996), 1.

59. Richard Dawkins, *Climbing Mount Improbable* (New York: W.W. Norton, 1996), 18.

60. Gould and Gould, *Animal Architects*, 55.

61. Gould and Gould, *Animal Architects*, 55.

62. Anaïs Khuong et al., "Stigmergic Construction and Topochemical Information Shape Ant Nest Architecture," *PNAS* 113, no. 5 (February 2, 2016): 1303–1308.

63. Perna and Theraulaz, "When Social Behaviour is Moulded in Clay," 89.

6. More Evolutionary Conundrums

1. Richard Dawkins, *The Blind Watchmaker* (New York: Norton, 1986), 112.

2. Scott Freeman and Jon C. Herron, eds., *Evolutionary Analysis*, 4th ed. (Upper Saddle River, NJ: Pearson–Prentice Hall, 2007), 802, "homoplasy." Other forms of homoplasy are parallel evolution and evolutionary reversal. Cladistics is a biological classification system.

3. Douglas Futuyma, *Evolution*, 3rd ed. (Sunderland, MA: Sinauer Associates, 2013), G-3.

4. "Convergence: Marsupials and Placentals," PBS Evolution Library, WGBH Educational Foundation and Clear Blue Sky Productions, 2001, https://www-tc.pbs.org/wgbh/evolution/library/01/4/pdf/l_014_02.pdf.

5. Richard Dawkins, *The Blind Watchmaker: Why the Evidence of Evolution Reveals a Universe Without Design* (New York: W. W. Norton, 1996), 94.

6. Ernst Mayr, *What Evolution Is* (New York: Basic Books, 2001), 224.

7. Brian K. Hall writes, "It is more inclusive and appropriate to speak of phenotypic homology to reflect the fact that other aspects of the phenotype, including behavior, can be recognized. Behavioral homology reminds us that phenotypic homology is much broader than structural and genetic homology." In "Homology, Homoplasy, Novelty, and Behavior," *Developmental Psychobiology* 55 (2013): 8, https://doi.org/10.1002/dev.21039.

8. Simon Conway Morris, *Life's Solution: Inevitable Humans in a Lonely Universe* (Cambridge: Cambridge University Press, 2005). Morris also has a website (www.mapoflife.org) that documents convergences, including a variety of types of behaviors.

9. Morris, *Life's Solution*, 285–286.

10. George McGhee, *Convergent Evolution: Limited Forms Most Beautiful* (Cambridge, MA: MIT Press, 2011), 7.

11. Edward O. Wilson, *Sociobiology: The New Synthesis* (Cambridge, MA: Belknap Press, 2000), 434.

12. Scott Gilbert interview by John Whitfield, in "Biological Theory: Postmodern Evolution?" *Nature* 455 (September 17, 2008): 281–284, https://doi.org/ https://doi.org/10.1038/455281a.

13. Charles Darwin, *On the Origin of Species: By Means of Natural Selection, or the Preservation of Favoured Races in the Struggle for Life* (London: John Murray, 1859), 193–194.

14. Mayr, *What Evolution Is*, 222.

15. Morris, *Life's Solution*, 283.

16. Morris, *Life's Solution*, 297.

17. Morris, *Life's Solution*, 205.

18. Sean B. Carroll, *Endless Forms Most Beautiful: The New Science of Evo-Devo* (New York: W. W. Norton, 2005).

19. A. L. Toth and G.E. Robinson, "Evo-Devo and the Evolution of Social Behavior," *Trends in Genetics* 23, no. 7 (2007): 334–341.

20. Rinaldo C. Bertossa, "Morphology and Behaviour: Functional Links in Development and Evolution," *Philosophical Transactions of the Royal Society of London B* 366, no. 1574 (2011): 2062.

21. Jerry A. Coyne, *Faith vs. Fact: Why Science and Religion are Incompatible* (New York: Penguin Publishing Group, 2015), 143.

22. McGhee, *Convergent Evolution*, 273.

23. Luc-Alain Giraldeau, "The Function of Behavior," in *The Behavior of Animals: Mechanisms, Function, and Evolution*, eds. Johan Bolhuis and Luc-Alain Giraldeau (Malden, MA: Blackwell Publishing, 2005), 200.

24. Mauricio Papini describes the importance of brain structures and algorithms: "Behavior is ultimately shaped by evolutionary forces, but proximally determined by the functioning of neural networks within the central nervous system of animals. It follows, then, that behavioral evolution is achieved by affecting neurological development." Mauricio R. Papini, *Comparative Psychology: Evolution and Development of Behavior* (New York: Psychology Press, 2008), 159.

25. Tristram D. Wyatt, *Pheromones and Animal Behavior: Chemical Signals and Signatures* (Cambridge University Press, Cambridge, 2014), 16.

26. Wyatt, *Pheromones and Animal Behavior*, 93.

27. Michael Lynch, "The Frailty of Adaptive Hypotheses for the Origins of Organismal Complexity," *PNAS* 104 (May 15, 2007): 8597, https://doi.org/10.1073/pnas.0702207104. See also Arlin Stolfuz, "Constructive Neutral Evolution: Exploring Evolutionary Theory's Curious Disconnect," *Biology Direct* 7 (2012): 35.

28. Andrew D. Kern and Matthew W. Hahn, "The Neutral Theory in Light of Natural Selection," *Molecular Biology and Evolution* 35, no. 6 (May 2018): 1366, https://doi.org/10.1093/molbev/msy092.

29. Jeffrey D. Jensen et al., "The Importance of the Neutral Theory in 1968 and 50 Years on: A Response to Kern and Hahn 2018," *Evolution* 73, no. 1 (January 2019): 111–114, https://doi.org/10.1111/evo.13650.

30. Jerry A. Coyne, *Why Evolution is True* (New York: Viking, 2009), 14, 134.

31. Darwin, *On the Origin of Species*, 270.

32. Darwin, *On the Origin of Species*, 183.

33. This phenomenon is discussed by Michael Behe in *Darwin Devolves: The New Science about DNA That Challenges Evolution* (New York: HarperOne, 2019).

34. Geoffrey K. Adams et al., "Neuroethology of Decision Making," in *Evolution and the Mechanisms of Decision Making*, eds. Peter Hammerstein and Jeffrey R. Stevens (Cambridge, MA: MIT Press, 2012), 84.

35. Adams et al., "Neuroethology of Decision Making," 85.

36. Eric R. Kandel, James H. Schwartz, and Thomas M. Jessell, eds., *Principles of Neural Science* (New York: McGraw-Hill, 2000), 1248–1250.

37. Irwin B. Levitan and Leonard K. Kaczmarek, *The Neuron* (Oxford: Oxford University Press, 2002), 553.

38. Janet L. Leonard and John P. Edstrom, "Parallel Processing in an Identified Neural Circuit: The *Aplysia californica* Gill-Withdrawal-Response Model System," *Biological Reviews* 79, no. 1 (February 2004): 42, https://doi.org/10.1017/S1464793103006183.

39. Leonard and Edstrom, "Parallel Processing," 51.

40. Adams et al., "Neuroethology of Decision Making," 86.

41. Jürgen Tautz, *The Buzz about Bees: Biology of a Superorganism* (Berlin: Springer-Verlag, 2008), 71.

42. Kenneth R. Miller, *Only a Theory: Evolution and the Battle for America's Soul* (New York: Penguin Books, 2008), 77–78.

43. Futuyma, *Evolution*, 285. Futuyma explains further, "A program likewise resides in a computer chip, but whereas that program has been shaped by an intelligent designer, the information in DNA has been shaped by a historical process of natural selection. Modern biology views the development, physiology, and behavior of organisms as the results of purely mechanical processes, resulting from interactions between programmed instructions and environmental conditions or triggers."

44. Dawkins, *The Blind Watchmaker* (996), 5.

45. Covering law theory, originally proposed by Carl Hempel, holds that "to give a scientific explanation of some phenomena, event or fact is to show how it can be seen to follow from a law (or set of laws) together with a specification of initial conditions." James Ladyman, *Understanding Philosophy of Science* (London: Routledge, 2002), 200.

46. Michael Ruse, "Methodological Naturalism Under Attack," in *Intelligent Design Creationism and its Critics*, ed. Robert Pennock (Cambridge: MIT Press, 2001), 365.

47. Elliott Sober, *Philosophy of Biology* (Boulder: Westview Press, 2000), 72.

48. Ernst Mayr, *Toward a New Philosophy of Biology* (New York: Basic Books, 2001), 19.

49. Sober, *Philosophy of Biology*, 73.

50. Manfred Eigen, *Steps Towards Life: A Perspective on Evolution* (Oxford: Oxford University Press, 1992), 12.

51. Eigen, *Steps Towards Life*, 13.

52. Eigen, *Steps Towards Life*, 17.

53. William Dembski, *No Free Lunch: Why Specified Complexity Cannot be Purchased Without Intelligence* (Lanham: Rowman and Littlefield Publishers, 2002), 152. Dembski writes, "Functional relationships at best preserve what information is already there, or else degrade it—they never add to it."

54. Stephen C. Meyer, *Signature in the Cell: DNA and the Evidence for Intelligent Design* (New York: HarperOne, 2009), 258.

55. Dembski, *No Free Lunch*, 155.

56. Douglas D. Axe, "Extreme Functional Sensitivity to Conservative Amino Acid Changes on Enzyme Exteriors," *Journal of Molecular Biology*, 301 (2000): 585–595; Douglas D. Axe, "Estimating the Prevalence of Protein Sequences Adopting Functional Enzyme Folds," *Journal of Molecular Biology*, 341 (2004): 1295–1315; Douglas D. Axe, "The Case Against a Darwinian Origin of Protein Folds," *BIO-Complexity* (2010):1–12, https://doi.org/10.5048/BIO-C.2010.1; Douglas D. Axe and Ann K. Gauger, "Model and Laboratory Demonstration That Evolutionary Optimization Works Well Only If Preceded by Invention—Selection Itself Is Not Inventive," *BIO-Complexity* (2015):1–13, https://doi.org/10.5048/BIO-C.2015.2.

57. See for example Dembski, *No Free Lunch*; William A. Dembski and Robert J. Marks II, "The Search for a Search: Measuring the Information Cost of Higher Level Search," *Journal of Advanced Computational Intelligence and Intelligent Informatics* 14 (2010): 475–486; Winston Ewert, George Montañez, William A. Dembski, and Robert J. Marks II, "Efficient Per Query Information Extraction from a Hamming Oracle," *Proceedings of the 42nd Meeting of the Southeastern Symposium on System Theory, IEEE, University of Texas at Tyler* (March 7–9, 2010): 290–297; Winston Ewert, William A. Dembski, and Robert J. Marks II, "Evolutionary Synthesis of Nand Logic: Dissecting a Digital Organism," *Proceedings of the 2009 IEEE International Conference on Systems, Man, and Cybernetics, San Antonio, TX* (October 2009): 3047–3053; William A. Dembski and Robert J. Marks II, "Bernoulli's Principle of Insufficient Reason and Conservation of Information in Computer Search," *Proceedings of the 2009 IEEE International Conference on Systems, Man, and Cybernetics San Antonio, TX* (October 2009): 2647–2652; William A. Dembski and Robert J. Marks II, "Conservation of Information in Search: Measuring the Cost of Success," *IEEE Transactions on Systems, Man and Cybernetics— Part A, Systems and Humans* 39, no. 5 (September 2009): 1051–1061.

58. Dembski, *No Free Lunch*, 149.

59. The program is called *EV*, developed by Thomas Schneider at the National Institutes of Health. Other evolution models include *Avida* and *Tierra*.

60. Miller, *Only a Theory*, 77.

61. Robert J. Marks II, William A. Dembski, and Winston Ewert, *Introduction to Evolutionary Informatics* (Singapore: World Scientific, 2017), 1.

62. Marks, Dembski, and Ewert, *Introduction to Evolutionary Informatics*, 187.

63. Marks, Dembski, and Ewert, *Introduction to Evolutionary Informatics*, 243.

64. Winston Ewert et al., "Time and Information in Evolution," *BIO-Complexity* (2012): 4, https://doi.org/10.5048/BIO-C.2012.4.

65. Marks, Dembski, and Ewert, *Introduction to Evolutionary Informatics*, 59.

66. Tom Strachan and Andrew P. Read, *Human Molecular Genetics 3* (London: Garland Science, 2004), 470. They further add, "The only mechanism that commonly generates novel functions is when a chromosomal rearrangement joins functional exons (DNA sequence) of two different genes. Such exon shuffling was no doubt important in evolution; for molecular pathology, it is most often noticed when it leads to cancer," 470.

67. The conclusion from one of the studies on the Lenski experiment was that "despite sustained adaptive evolution in the long-term experiment, the signature of selection is much weaker than that of mutational biases in mutator genomes." Alejandro Couce et al., "Mutator Genomes Decay, Despite Sustained Fitness Gains, in a Long-Term Experiment

with Bacteria," *PNAS* 114, no. 43 (October 10, 2017): E9026, https://doi.org/10.1073/pnas.1705887114.

68. Behe, *Darwin Devolves*, 256. See also Behe, *The Edge of Evolution: The Search for the Limits of Darwinism* (New York: Free Press, 2007).

69. Manolis Kellis et al., "Defining Functional DNA Elements in the Human Genome," *PNAS* 111, no. 17 (April 2014): 6131, https://doi.org/10.1073/pnas.1318948111.

70. The ENCODE Project Consortium, "An Integrated Encyclopedia of DNA Elements in the Human Genome," *Nature* 489 (September 2012): 57–74, https://doi.org/10.1038/nature11247.

71. Kellis et al., "Defining Functional DNA Elements," 6131.

72. Benjamin E. R. Rubin et al., "Rate Variation in the Evolution of Non-Coding DNA Associated with Social Evolution in Bees," *Philosophical Transactions of the Royal Society B* 374, no. 1777 (July 2019): 20180247, https://doi.org/10.1098/rstb.2018.0247.

73. Robert Stalnaker, *Inquiry* (Cambridge, MA: MIT Press, 1984), 85, quoted in Dembski, *No Free Lunch*, 125.

74. Robert K. Barnhart, *Hammond Barnhart Dictionary of Science* (Maplewood, NJ: Hammond, 1986).

75. Marcel P. Schüzenberger, "Algorithms and the Neo-Darwinian Theory of Evolution," in *Mathematical Challenges to the Neo-Darwinian Interpretation of Evolution*, eds. Paul S. Moorehead and Martin M. Kaplan (Wistar Institute Press, 1966), 74–75.

76. See the examples discussed by Behe in *Darwin Devolves*.

77. The Law of Conservation of Information states that "the complex specified information in an isolated system of natural causes does not increase." Dembski, *No Free Lunch*, 169.

78. Marks II, Dembski, and Ewert, *Evolutionary Informatics*.

7. Complex Programmed Behaviors— Intelligently Designed

1. William A. Dembski, *The Design Revolution: Answering the Toughest Questions about Intelligent Design* (Downers Grove, IL: InterVarsity Press, 2004), 27.

2. See discussions of this issue in James Ladyman, *Understanding Philosophy of Science* (London: Routledge, 2008), and Peter Godfrey-Smith, *Theory and Reality* (Chicago: University of Chicago Press, 2003).

3. Peter Lipton, *Inference to the Best Explanation* (London: Routledge, 2005).

4. Ted Honderich, ed., *The Oxford Guide to Philosophy* (Oxford: Oxford University Press, 2005), s.v. "induction."

5. Richard H. Popkin and Avrum Stroll, *Philosophy Made Simple* (New York: Made Simple Books, 1993), 240.

6. James Ladyman explains that the result of applying this process is, "Once we have inductively inferred our generalization in accordance with the scientific method, then it assumes the status of a law or theory and we can use deduction to deduce consequences of the law that will be predictions or explanations." Ladyman, *Understanding Philosophy of Science*, 29.

7. Lipton further writes, "Inductive inference is thus a matter of weighing evidence and judging probability, not of proof." Lipton, *Inference to the Best Explanation*, 5.

8. Lipton, *Inference to the Best Explanation*, 58.

9. *Stanford Encyclopedia of Philosophy*, s.v. "abduction," accessed May 18, 2021, https://plato. stanford.edu/entries/abduction/.

10. Charles Darwin, *The Origin of Species: By Means of Natural Selection, or the Preservation of Favoured Races in the Struggle for Life*, 6th ed. (London: John Murray, 1872), 421, http:// darwin-online.org.uk/content/frameset?itemID=F391&viewtype=image&pageseq=1.

11. Günter Bechly and Stephen Meyer, "The Fossil Record and Universal Common Ancestry," *Theistic Evolution: A Scientific, Philosophical, and Theological Critique* (Wheaton, Illinois: Crossway, 2017), 331.

12. Doori Park et al., "Uncovering the Novel Characteristics of Asian Honey Bee, Apis cerana, by Whole Genome Sequencing," *BMC Genomics* 16, no. 1 (January 2015): 5, https:// doi.org/10.1186/1471-2164-16-1.

13. Daniel F. Simola et al., "Social Insect Genomes Exhibit Dramatic Evolution in Gene Composition and Regulation While Preserving Regulatory Features Linked to Sociality," *Genome Research* 23 (2013): 1235.

14. David Snoke, "Systems Biology as a Research Program for Intelligent Design," *BIO-Complexity* 3 (2014):1–11, https://doi.org/10.5048/BIO-C.2014.3.

15. Snoke, "Systems Biology."

16. *Merriam-Webster*, s.v. "biomimetics," accessed August 20, 2021, https://www.merriam-webster.com.

17. Jangsun Hwang et al., "Biomimetrics: Forecasting the Future of Science, Engineering, and Medicine," *International Journal of Nanomedicine* 10 (2015): 5705.

18. R. Haven Wiley, *Noise Matters: The Evolution of Communication* (Cambridge, MA: Harvard University Press, 2015), 128.

19. Alejandro F. Villaverde and Julio R. Banga, "Reverse Engineering and Identification in Systems Biology: Strategies, Perspectives and Challenges," *Journal of the Royal Society Interface* 11, no. 91 (February 2014): 20130505, https://doi.org/10.1098/rsif.2013.0505.

20. Robert C. Richardson, "Engineering Design and Adaptation," *Philosophy of Science* 70 (December 2003): 1277.

21. William A. Dembski, *Intelligent Design: The Bridge Between Science and Theology* (Downers Grove, IL: InterVarsity Press, 1999); *The Design Revolution: Answering the Toughest Questions about Intelligent Design* (Downers Grove, IL: InterVarsity Press, 2004), 317.

22. George M. Church, Yuan Gao, and Sriram Kosuri, "Next-Generation Digital Information Storage in DNA," *Science* 337 (September 2012): 1628.

23. Yaniv Erlich and Dina Zielinski, "DNA Fountain Enables a Robust and Efficient Storage Architecture," *Science* 355 (March 2017): 950.

24. Benjamin E. R. Rubin et al., "Rate Variation in the Evolution of Non-Coding DNA Associated with Social Evolution in Bees," *Philosophical Transactions of the Royal Society B* 374, no. 1777 (July 2019): 20180247, https://doi.org/10.1098/rstb.2018.0247.

25. *Oxford English Dictionary* (Oxford: Clarendon Press, 1989), s.v. "teleology."

26. Michael Ruse, *Darwin and Design* (Cambridge, Harvard University Press, 2003), 126.

27. Francisco Ayala, "Design Without Designer," in *Debating Design*, eds. William Dembski and Michael Ruse (New York: Cambridge University Press, 2004), 66.

28. Jerry A. Coyne, *Faith vs. Fact: Why Science and Religion are Incompatible* (New York: Penguin Books, 2015), 19.

29. Oxford English Dictionary, s.v. "teleophobia."

30. Coyne, *Faith vs. Fact*, 216.

31. Charles Darwin, *On the Origin of Species: By Means of Natural Selection, or the Preservation of Favoured Races in the Struggle for Life* (London: John Murray, 1859), 490, https://archive.org/details/onoriginspeciesf00darw/page/n501/mode/2up.

32. Michael J. Behe, *The Edge of Evolution: The Search for the Limits of Darwinism* (New York: Free Press, 2007), 142.

8. Answering Common Objections to Intelligent Design

1. Vincent de Paul to a Priest of the Mission in Saintes, December 28, 1650, Letter no. 1301, *Saint Vincent de Paul Correspondence, Conferences, Documents, 1581-1660*, vol. 4, *April 1650–July 1653* [1921], ed. and trans. Marie Poole (New Rochelle, NY: New City Press, 1993), 132, https://via.library.depaul.edu/vincentian_ebooks/29/.

2. Michael Behe, William Dembski, Stephen Meyer, and others have addressed the common objections to intelligent design theory ably and at length. Stephen Meyer addresses several objections in chapter 18 of *Signature in the Cell: DNA and the Evidence for Intelligent Design* (New York: HarperOne, 2009). Michael Behe answered many of his critics in *A Mousetrap for Darwin* (Seattle: Discovery Institute Press, 2020). See also the criticisms and responses in William Dembski and Michael Ruse, eds., *Debating Design* (New York: Cambridge University Press, 2004).

3. *Oxford Pocket Dictionary and Thesaurus: American Edition* (New York: Oxford University Press, 1997), s.v. "naturalism." Ontology refers to the nature of being.

4. Bruce L. Gordon, "A Quantum-Theoretic Argument Against Naturalism," in *The Nature of Nature: Examining the Role of Naturalism in Science*, eds. Bruce Gordon and William Dembski (Wilmington, DE: ISI Books, 2011), 179. Gordon further explains that "the philosophical naturalist insists on the causal closure of the material realm. A corollary of this viewpoint is that there is no such being as God or anything remotely resembling him; rather, according to the naturalist, the spatio-temporal universe of our experience, in which we exist as strictly material beings, is causally self-sufficient. The explanatory resources of this naturalistic metaphysical closure are restricted, therefore, to material objects, causes, events, and processes and their causally emergent properties," 180.

5. Sagan repeats this in a book that was based on the television series. Carl Sagan, *Cosmos* (New York: Random House, 1980), 4.

6. Ernan McMullin, "Varieties of Methodological Naturalism," in *The Nature of Nature*, 83.

7. McMullin, "Varieties of Methodological Naturalism," 82.

8. Michael Rea, *World Without Design: The Ontological Consequences of Naturalism* (Oxford: Oxford University Press, 2004), 52. Rea writes, "If everything else is at the mercy of science, why not naturalism? So long as scientific method remains intact as a way of judging between two theses, naturalism will always prescribe taking sides with science and could therefore never find itself condemned by science."

9. "Bibliographic and Annotated List of Peer-Reviewed Publications Supporting Intelligent Design," Discovery Institute Center for Science and Culture, July 2017, https://www.discovery.org/m/2018/12/ID-Peer-Review-July-2017.pdf.

10. *Science, Evolution, and Creationism* (Washington, DC: National Academy of Sciences, 2008), 42–43.

11. Stephen C. Meyer, *Darwin's Doubt: The Explosive Origin of Animal Life and the Case for Intelligent Design* (New York: HarperOne, 2013), 343. Meyer also has a detailed explanation of this method, which he describes as "abductive inference."

12. Stephen C. Meyer, "Sauce for the Goose," in *The Nature of Nature*, 100.

13. Daniel Dennett, "Show Me the Science," *New York Times*, August 28, 2005.

14. Michael J. Behe, *The Edge of Evolution: The Search for the Limits of Darwinism* (New York: Free Press, 2007), 142.

15. Dustin J. Van Hofwegen, Carolyn J. Hovde, and Scott A. Minnich, "Rapid Evolution of Citrate Utilization by *Escherichia coli* by Direct Selection Requires citT and dctA," *Journal of Bacteriology* 198, no. 7 (April 2016): 1022–1034.

16. Jerry Fodor and Massimo Piattelli-Palmarini, *What Darwin Got Wrong* (New York: Farrar, Straus and Giroux, 2010), 213, note 5 to Chapter 7.

17. Meyer, "Sauce for the Goose," 111.

18. Peter Lipton, *Inference to the Best Explanation* (London: Routledge, 2004), 164.

19. Lipton, *Inference to the Best Explanation*, 165.

20. Meyer, "Sauce for the Goose," 100.

21. Meyer, *Signature in the Cell*, 482.

22. Kenneth R. Miller, *Only a Theory: Evolution and the Battle for America's Soul* (New York: Penguin Books, 2008), 168.

23. Miller, *Only a Theory*, 192.

24. Douglas Futuyma, *Evolution*, 3rd ed. (Sunderland, MA: Sinauer Associates, Inc., 2013), 285.

25. Thomas Henry Huxley, *Darwiniana: Essays, Volume 2* [1893] (Scotts Valley, CA: CreateSpace Independent Publishing Platform, 2015), loc. 836, Kindle.

26. Adam Arkin and John Doyle, "Appreciation of the Machinations of the Blind Watchmaker," *IEEE Special Issue on Systems Biology* 53 (January 2008): 8–9, https://doi.org/10.1109/TAC.2007.913342.

27. Innes Cuthill, "The Study of Function in Behavioral Ecology," in *Tinbergen's Legacy: Function and Mechanisms in Behavioral Biology*, eds. Johan Bolhuis and Simon Verhulst (Cambridge: Cambridge University Press, 2009), 111.

28. Luc-Alain Giraldeau, "The Function of Behavior," in *The Behavior of Animals: Mechanisms, Function, and Evolution*, eds. Johan Bolhuis and Luc-Alain Giraldeau (Malden, MA: Blackwell Publishing, 2005), 199.

29. Tim Lewens, *Organisms and Artifacts: Design in Nature and Elsewhere* (Cambridge: The MIT Press, 2004), 2.

30. Lewens, *Organisms and Artifacts*, 2.

31. Lewens calls the neo-Darwinian view the "selected effects" account: "The function of a trait T is F iff (if and only if) T was selected for F." He calls the version that would be identified as the ID view the "intended effects" account, which he defines as, "The function of artifact A is F iff some agent X intends that A perform F." Lewens, *Organisms and Artifacts*, 90.

32. J. Scott Turner makes a strong argument for the purposefulness in mechanisms such as homeostasis in *Purpose and Desire: What Makes Something "Alive" and Why Modern Darwinism Has Failed to Explain It* (New York: HarperOne, 2017).

33. Hayne W. Reese, "Teleology and Teleonomy in Behavior Analysis," *The Behavior Analyst* 17, no. 1 (Spring 1994): 88.

34. Reese, "Teleology and Teleonomy in Behavior Analysis." Reese explains why "selection by consequences and some other principles in behavior analysis are teleological only in a peculiar sense of teleology, and when these principles are called teleological, readers and listeners may well misunderstand," 89.

35. Mary Jane West-Eberhard, *Developmental Plasticity and Evolution* (Oxford: Oxford University Press, 2003), 314.

36. Ernst Mayr, *Toward a New Philosophy of Biology: Observations of an Evolutionist* (Cambridge: Harvard University Press, 1988), 45.

37. An example is the following: "The argument from design thus focused attention onto the adaptation of structure to function. God is not only wise, he is also benevolent because he gives each species exactly what it needs to live in the place where he created it. The argument presupposes a static creation, in which species and their environments remain just as they were when first created. It has often been said that Darwin would turn the argument from design on its head by showing that adaptation is a process by which species are adjusted to changing environments." Peter Bowler and Iwan Rhys Morus, *Making Modern Science* (Chicago: University of Chicago Press, 2005), 132.

38. Miller, *Only a Theory*, 124.

39. Many of Gould's works underscore the point that the fossil record strongly testifies that stasis is the norm in the history of life. For one example see *The Structure of Evolutionary Theory* (Cambridge: Belknap Press of Harvard University Press, 2002), where he discusses stasis at numerous points in the text.

40. Mayr, *Toward a New Philosophy of Biology*, 151.

41. Daniel Dennett, *Darwin's Dangerous Idea: Evolution and the Meanings of Life* (New York: Touchstone, 1995), 216.

42. Francisco Ayala, "Design Without Designer," in *Debating Design*, 56.

43. For accounts of the flawed arguments against the design of the vertebrate eye, see Casey Luskin, "Eyeballing Design," *Salvo* (Winter 2011), 54–56; Hallie Kemper, Gary Kemper, and Casey Luskin, *Discovering Intelligent Design: A Journey Into the Scientific Evidence* (Seattle, WA; Discovery Institute Press, 2013), 132–133; and Jonathan Wells, *Zombie Science: More Icons of Evolution* (Seattle, WA: Discovery Institute Press, 2017), 142–146.

44. Charles Darwin to Asa Gray, May 22, 1860, Darwin Correspondence Project, Letter no. 2814, University of Cambridge, https://www.darwinproject.ac.uk/letter/DCP-LETT-2814.xml.

45. Charles Darwin to J. D. Hooker, July 13, 1856, Darwin Correspondence Project, Letter no. 1924, https://www.darwinproject.ac.uk/letter/DCP-LETT-1924.xml

46. Richard Dawkins, *River Out of Eden: A Darwinian View of Life* (New York: Basic Books, 1995), 132. A page later Dawkins doubles down, commenting, "The universe we observe has precisely the properties we should expect if there is, at bottom, no design, no purpose, no evil and no good, nothing but blind pitiless indifference."

47. Michael Ruse, *Darwin and Design* (Cambridge: Harvard University Press, 2003), 331.

48. Fritz Trillmich, "Parental Care: Adjustments to Conflict and Cooperation," in *Animal Behaviour: Evolution and Mechanisms*, ed. Peter Kappeler (Heidelberg: Springer-Verlag, 2010), 285.

49. Nicholas Davies, John Krebs, and Stuart West, *An Introduction to Behavioural Ecology* (Oxford: Wiley-Blackwell, 2012), 4.

50. Davies, Krebs, and West, *An Introduction to Behavioural Ecology*, 232.

51. Douglas W. Mock, Hugh Drummond, and Christopher H. Stinson, "Avian Siblicide," in *Exploring Animal Behavior*, eds. Paul W. Sherman and John Alcock (Sunderland, MA: Sinauer Associates, 2010), 232–243.

52. Mock, Drummond, and Stinson, "Avian Siblicide," 235.

53. Mock, Drummond, and Stinson, "Avian Siblicide," 239–240.

54. John Alcock, *Animal Behavior: An Evolutionary Approach* (Sunderland, MA: Sinauer Associates, 2013), 9.

55. Alcock, *Animal Behavior*, 10.

56. Trillmich, "Parental Care: Adjustments to Conflict and Cooperation," 279.

57. Joan E. Strassmann and David C. Queller, "Insect Societies as Divided Organisms: The Complexities of Purpose and Cross-Purpose," in *In the Light of Evolution*, eds. John C. Avise and Francisco J. Ayala, vol. 1, *Adaptation and Complex Design* (Washington: National Academies Press, 2007), 159.

58. David Hull, "The God of the Galapagos," *Nature* 352 (1991): 485–86, https://doi.org/10.1038/352485a0.

59. Cornelius G. Hunter, *Darwin's God: Evolution and the Problem of Evil* (Grand Rapids: Brazos Press, 2001), 11.

60. Hunter, *Darwin's God*, 16.

61. Hunter, *Darwin's God*, 16.

62. Since the objection is theological in nature, it would be illogical to object to an answer that draws upon theological resources. One possible theological explanation is provided by Christian theology and the idea of "the Fall" and entrance of sin into the world. In this theology the appearance of sin results in death, disease, and other maladies. Michael J. Murray has an excellent discussion of this in *Nature Red in Tooth and Claw* (Oxford: Oxford University Press, 2011).

63. Richard Dawkins, *The God Delusion* (Boston: Houghton Mifflin Co., 2006), 157–58.

64. Dennett, *Darwin's Dangerous Idea*, 71.

65. William Lane Craig, *Reasonable Faith* (Wheaton: Crossway Books, 2008), 171.

66. Michael Augros, *Who Designed the Designer?* (San Francisco: Ignatius Press, 2015).

FIGURE CREDITS

Figure 1.1. Monarch butterflies. "Selective Focus Photography of Group of Monarch Butterflies Perching on Purple Lavender Flower." Photograph by Cindy Gustafson, 2017, Pexels. Pexels license.

Figure 2.1. Arctic tern. "4848965." Photograph by Paul Williams (Paulw61), 2020, Pixabay. Pixabay license.

Figure 3.2. Loggerhead sea turtle. "Loggerhead turtle, Caretta caretta, in open water." Photograph by Jon, undated, Adobe Stock. Adobe license.

Figure 4.2. Leafcutter ants at work. "Leaf cutter ants." Photograph by Adrian Pingstone, March 2008, Wikimedia Commons. Public domain.

Figure 5.1. Weaver ant nest construction. "Nest Construction by Oecophylla smaragdine Workers, Thailand." Photograph by Sean Hoyland, February 24, 2008, Wikimedia Commons. Public domain.

Figure 5.2. Cathedral termite mound. "Cathedral Termite Mound." Photograph by J. Brew, April 30, 2009, Wikipedia. CC-BY-SA 2.0 license.

INDEX

Printed in Great Britain
by Amazon

77898487R00141